"This book gently leads the non-specialist into the wonderful world of symmetry . . ."

–H.S.M. Coxeter
Professor of Mathematics
University of Toronto, Canada
author of *Regular Polytopes.*

"In this richly illustrated book, the Hargittais bring us a captivating, compelling view of the patterns in the familiar and the obscure."

–Roald Hoffmann
Chemistry Nobel laureate
Cornell University
Ithaca, NY

"By the variety and excellence of the photographs, as well as the text, brief but to the point, the authors have produced a marvelous book revealing the aspects of symmetry in all its varied forms."

–Herbert Hauptman
Chemistry Nobel laureate
Hauptman-Woodward Medical
Research Institute, Buffalo, NY

"This exploration ranges from the simple mirror image, images reflected and rotated in planes and the mathematical descriptions for these and other models. The added extra is the enormous number of images that accompany the simple text . . ."

–Maggie McDonald, *New Scientist*

"This 221-page, profusely illustrated compendium is a pleasure to browse through, yet fully repays the reader who stops to dive in more deeply."

–*Quantum*

"A unique look at life and creation."

–*Booklist*

"This book is the first step in understanding our world in a scientific and materialist way."

–Alan L. Mackay, F.R.S.
Birkbeck College
London, U.K.

"From one category to another, from image to image as from one dot to the next, our visual memory moves in often unconnected segments. In their beautiful book the Hargittais trace the unifying line that connects the dots, and the reader experiences the thrill of discovery and deep reverence for beauty without which the book itself would not have been possible."

–Gabriela Radulescu
Scipress
Chicago, IL

"Why leave mathematics in the classroom when the Hargittais will show it on Coke machines, soldiers, movie posters, parked buses, bank logos, athletes, crystals and flowers?"

–Steve Baer
Zomeworks Corp.,
Albuquerque, NM

"An . . . incredible compilation."

–*The Mathematics Teacher*

SYMMETRY
A UNIFYING CONCEPT

Isttexttextván Hargittai
Magdolna Hargittai

Shelter Publications, Inc., Bolinas, California

Distributed in the United States by Random House, Inc., New York, N.Y. and in Canada by Random House of Canada Limited, Toronto

Library of Congress Cataloging in Publication Data

Hargittai, István.
    Symmetry: a unifying concept / István Hargittai, Magdolna Hargittai.
        p.   cm.
    Includes bibliographical references and index.
    ISBN 0-936070-17-X (Shelter Publications)
    ISBN 0-679-76945-5 (Random House)
    1. Symmetry. 2. Symmetry (Physics) I. Hargittai, Magdolna. II. Title.
    Q172.5.S95H37   1994   93-42632
    516'.1—dc20   CIP

2 3 4 5 6 — 99 98 97 96
Lowest numbers indicate number and year of this printing.

First Printing: April, 1994

Printed in the United States of America

Additional copies of this book may be purchased for $18 plus $3 shipping and handling from:

Shelter Publications, Inc.
P.O. Box 279
Bolinas, California 94924 USA

Or write for a free catalog of books.

FRONT COVER PHOTO: *Bridge over Tiber River in Rome*

REAR COVER PHOTO: *Spiral staircase in a building on the campus of the Norwegian Technical University in Trondheim, Norway*

TITLE PAGE PHOTO: *The Washington Monument and the Reflecting Pool, Washington, D.C.*

NOTE: This book, when held open, forms a golden rectangle *(see p.161).*

To Balázs and Eszter

CONTENTS

INTRODUCTION

Symmetry is a word used often in everyday language, and we all recognize symmetry when we see it. The human body, butterflies, flowers, animals, buildings, decorations, artistic creations and countless other natural or human-made objects are symmetrical.

In the sciences, symmetry is frequently used as a technical term. Unlike other technical words, however, where there is an appreciable difference in meaning to scientists and the general public, symmetry means more or less the same thing to both technical and nontechnical people alike.

But what is symmetry, really? As with other fundamental concepts, it is not easy to provide a simple definition.

The Cathedral in Milan, Italy

Geometrical Symmetry

When something is symmetrical in the geometric sense, there is a one-to-one correspondence among its parts. The word itself comes from the Greek *symmetria*, meaning "the same measure." So, for example, the two sides of the human face show a one-to-one correspondence, as do all the parts of a five-petal flower or a five-spoke hubcap.

FAR RIGHT: *Henri Matisse,* Woman's Portrait

There are other symmetries that do not come so readily to mind when we think of symmetry. For example, you may not think of the railroad tracks at right as being symmetrical. Yet there is a one-to-one correspondence among the parts and, accordingly, they are symmetrical.

The examples of symmetry we have mentioned so far are based on geometrical correspondence—symmetry is generated by reflection (face), rotation (flower and hubcap) or repetition (railroad tracks).

Fuzzy Symmetries

But what happens when things are not precise geometrically? We do not have to look closely in the mirror to see that the symmetry of the human face is not perfect. A close look at anything created in nature reveals that all symmetries have some imperfections. If we are willing to overlook small differences, we can "see" symmetry where it would not exist according to more rigorous geometrical criteria.

For example, above is a virtually perfect reflection of an ancient Roman bridge in the river Tiber. But reflection is far less perfect in these other photos of a small building materials plant and its reflection in the river. Here a mild breeze is rippling the water's surface and distorting the reflection. A perfect reflection would appear only under conditions of absolute stillness. Yet we can easily relate this impressionistic image to the original.

Our Ability to Geometrize

The human eye and mind have a remarkable ability to discern patterns or characteristic shapes even when there are irregularities or omissions in an overall pattern. Look, for example, at the beautiful medieval pattern of this Portuguese stamp.

Although the design is quite complete, some of the corners, especially the lower left edge, are damaged. Yet as we look at the stamp, we unconsciously skip over the damage and see the whole pattern as if it were perfect.

RIGHT: *An unusual example of multiple reflection: the pagoda and its reflection appear in both real life and in the young artist's painting.*

Or consider what appears to be a worn piece of wood with some unintelligible writing on its sides. It looks as if it were found on the banks of a river. It is, in fact, a minisculpture by a contemporary Swiss artist. Geometrically speaking, this object would not qualify as being symmetrical (or as being a cube, for that matter). Yet we have no difficulty in recognizing it as such.

Harmony and Proportion

Thus far we have discussed not only geometrical symmetry, but also how strict geometrical criteria can be relaxed. Beyond geometrical definitions, though, there is another, broader meaning to symmetry—one that relates to harmony and proportion, and ultimately to beauty. This aspect involves feeling and subjective judgment and, as a result, is especially difficult to describe in technical terms.

In fact, much of what you will see in this book has to do with the beauty and harmony we have discovered in our travels throughout the world. We have taken photos and utilized graphic material that not only conform to one or more defined symmetry principles, but that have often appealed to our aesthetic sensibilities as well.

A Unifying Concept

When all these materials are assembled in a book (or in one's mind, for that matter), a fascinating theme emerges: symmetry is a unifying concept.

Human fields of study, especially in modern times, have become increasingly compartmentalized. This is especially true in education. The sciences, the humanities, and the arts have all drifted apart over the years. There has also been an increasing trend toward separation (or specialization) within the scientific world itself: physics, chemistry, biology, etc.

Symmetry, however, can provide a connecting link. It is a unique tool for reuniting seemingly disparate fields of endeavor. Accordingly, symmetry can provide insight into what has been lost in the separations.

And considerations of harmony and proportion further help us to relate things that at first glance may appear to have no common ground at all.

The bridging ability of the symmetry concept is a powerful tool—it provides a perspective from which we can see our world as an integrated whole.

Sculpture, Simétria, *in the Prado, Madrid, Spain*

Two Major Symmetry Classes

There are many kinds of symmetry, but most of them can be divided into two large classes: point groups or space groups.

Point Groups

Here, the identifying factor is that at least one special point in the object or pattern differs from all the others. This special point (also called unique point) has an important distinguishing feature: it remains unchanged no matter what type of symmetry operation is performed. Such symmetries belong to the point-group category.

A circle, for example, has at its center a unique point, as you can see in the black and white cobblestone pattern in this Italian piazza. There is no other point equivalent to the center in the entire pattern. If we rotate this circular pattern around its center, the pattern remains unchanged, regardless of its position.

This is point-group symmetry because:

- there is one point, the center, which is unique
- that point is not repeated elsewhere in the pattern
- the point does not change during rotation

Let's consider another example. The human face has bilateral symmetry. If you look at the Matisse painting on page xii, you will see that all points along the line dividing the face vertically are unique points—the center of the forehead, the tip of the nose, the midpoints of the lips, etc. When you reflect one side of a face to produce a mirror image, these unique points stay in place. Thus reflection is another type of point-group symmetry.

Some simple drawings can further illustrate the difference between point groups and space groups.

The symmetries of a single cube belong to the point groups, since the cube has a unique point—its center.

If we repeat this cube in an endless row, we get a one-dimensional space group:

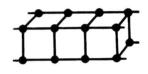

Space Groups

In the other class of symmetries, there is no special point in the object or pattern that is different from all the others. These are the space groups. In this Italian pavement, for instance, the pattern is created by a (seemingly) endless repetition of an arc.

Here the arcs in the pattern extend in two directions—length and width. The corresponding symmetry is called a two-dimensional space group. In general, space groups can be one-dimensional, two-dimensional or three-dimensional—according to whether the repetition extends in one, two, or three directions.

Repeating the cube in an endless plane, we get a two-dimensional space group:

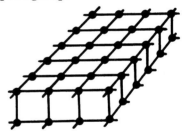

And, finally, repeating it in space by stacking cubes up in an endless structure, we get a three-dimensional space group:

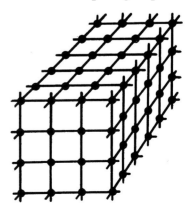

Organization of This Book

The structure of the book is based on the two main symmetry classes just described.

- **The first half** (chapters I–IX) deals with point-group symmetries.

- **The second half** (chapters XI–XV) covers space-group symmetries.

- **The middle** (chapter X) is quite special. It is about the symmetry of opposites, or antisymmetry, where geometrical symmetry is combined with color changes or other property reversals.

This is primarily a visual book. Because so many things related to symmetry have visual appeal, we have used photos, drawings, and paintings to illustrate basic symmetry concepts. The language is simple. We want everyone to be able to understand this book, so we have made a conscious effort to avoid technical terms. However, where specific recognized principles of symmetry are first introduced (such as symmetry **operations** or **elements**, we will use boxes to define the terms. *(See pp. 2, 39, 124, 131, for example.)*

The single, most important purpose of this book is to help you notice the world around you, to train your eye and mind to see new patterns and make new connections. A few years ago, soon after we had presented a slide show on symmetry, someone came up to us and said, with genuine anger, "You and your silly symmetries. I can't help seeing them everywhere since your talk."

We couldn't have been more delighted!

István Hargittai

Magdolna Hargittai

Mirror Symmetry

Bilateral symmetry is the symmetry everybody is aware of, and to many people this is symmetry itself. Bilateral symmetry occurs when two halves of a whole are each other's mirror images. Accordingly, bilateral symmetry is also called *mirror symmetry*.

Let us define some terms that you will see at various places throughout the book:

The *action* that characterizes a particular type of symmetry is called a **symmetry operation**. For example, using a mirror to make a whole from one of the halves of an object is a symmetry operation.

The means *(tool)* whereby the operation is performed is called a **symmetry element.** For example, the symmetry element here is a mirror.

We will use boxes, as below, to define **symmetry types**, **operations**, and **elements**, as well as other important terms and symbols.

Symmetry type
Bilateral symmetry:
Two halves of the whole are each other's mirror images

Symmetry operation (action)
Reflection:
Reflecting one-half of an object reconstructs the image of the whole object

Symmetry element (tool)
Mirror plane:
Also called reflection plane or symmetry plane. Applying a mirror plane to either of two halves, the whole is recreated

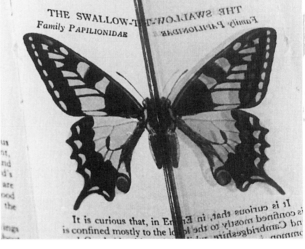

The left half of the butterfly and its mirror image depict the whole butterfly

The right half of the butterfly and its mirror image also depict the whole butterfly; therefore the butterfly has mirror symmetry

If the two halves are not mirror images of one another, the application of a mirror does not recreate the original object or figure. For example, the letter R does not have mirror symmetry.

The letter A, however, does have mirror symmetry.

We don't actually have to use a mirror to determine whether or not something has mirror symmetry. Just imagining a mirror in place between the left and right halves and envisioning the result is enough. Examples of bilateral symmetry abound in our daily lives. Moreover, this form of symmetry has been utilized repeatedly in the fine arts, as we shall see throughout the following pages.

Tyger! Tyger! burning bright
In the forests of the night,
What immortal hand or eye
Could frame thy fearful symmetry?

William Blake
"The Tyger"

Looking at these beautiful tigers and reading Blake's poem, we can understand that symmetry in this context means more than just the geometrical exactitude of the left and right sides of the tiger—here it is equivalent to beauty and harmony.

In Plants

The symmetry of plants is diverse. There are, however, some flowers, such as orchids, that have only mirror symmetry. So far, some 25,000 kinds of orchids have been described and they all exhibit mirror symmetry. Here are some examples from Oahu, Hawaii.

If you look at these orchids, you can see their bilateral symmetry. Put a mirror down the middle to reflect one side and the original image will be recreated.

The arrangement of leaves on stems frequently exhibits mirror symmetry.

The Human Body

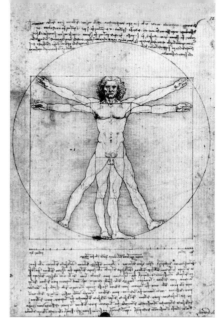

Leonardo da Vinci's famous drawing displays the proportions of the human body and stresses its bilateral symmetry.

Leonardo da Vinci, Schema-delle proporzioni del corpo umano

How bewitching the beauty of the human body, composed not of paint or stone, but of living, corruptible matter, charged with the secret fevers of life and decay! Consider the wonderful asymmetry of this structure: Shoulders and hips and nipples swelling on either side of the breast, and ribs arranged in pairs, and the navel centered in the belly's softness, and the dark sex between the thighs. Consider the shoulder blades moving beneath the silky skin of the back, and the backbone in its descent to the paired richness of the cool buttocks, and the great branching of vessels and nerves that passes from the torso to the arms by way of the armpits, and how the structure of the arms corresponds to that of the legs!

Thomas Mann
The Magic Mountain

The symmetry of the human body is clearly illustrated by the sculptures at right.

LEFT: *Standing Buddha, China, 18th century*

RIGHT: *King Mykerinos with goddess Hathor and nome goddess, Egypt, 2720 B.C.*

Mobility does not interfere with the bilateral symmetry of the human body. If anything, it even gives emphasis, as shown here by various athletes. The movements of gymnasts, divers and swimmers follow strict rules, and the perfection of their movements approximates geometrical symmetry.

Faces of defending guards, Buddhist temple, Korea

The Human Face

The human face contains mirror symmetry. However, sometimes there are minute variations on the two sides of the face which can be quite conspicuous. Portrait artists may render a face so that it appears more symmetrical than it is in reality in order to "idealize" the subject or to please the person who commissioned the work.

Sculpture of a Hungarian king

Egyptian sculpture

Aztec stone sculpture

St. Peter on St. Peter's Square, Vatican City

7

I. BILATERAL SYMMETRY

Notice the characteristic bilateral symmetry in these Native American masks and sculptures.

Kwakiutl mask

*Dancing mask,
Bellabella Indians of
British Columbia*

*Dancing mask,
Bellabella Indians of
British Columbia*

Eskimo mask

*Dancing mask shaped like
bird's head, Bellabella Indians
of British Columbia*

*Dancing mask,
Bellabella Indians of
British Columbia*

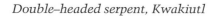

Double–headed serpent, Kwakiutl

Some artists stress the mirror symmetry of the human face by using the reflection plane.

There have been speculations that the right side of the human face is more "public," while the left side is more "private." Others have argued that the right side is more representative of the whole face than the left side.

LEFT: *Jenö Barcsay,* Woman's head

LEFT: *George Buday,*
Miklós Radnóti, *woodcut*

RIGHT: *Pablo Picasso,* Woman's head

Right	*Left*	*Right*	*Right*	*Left*	*Left*
(a)		*(b)*		*(c)*	

Compare the two sides of your face by standing in front of a mirror or shop window. Are there any differences?

(a) The real face of the poet Edgar Allan Poe. *(b)* The right side of Poe's face with its own reflection. *(c)* The left side of the same face with its own reflection. Pictures *(b)* and *(c)* are strikingly different, emphasizing as they do the differences between the left and right sides of the poet's face.

Double Heads

This double-headed dog was drawn merely to attract attention in a Belgian ad.

St. Petersburg, Russia

Zürich, Switzerland

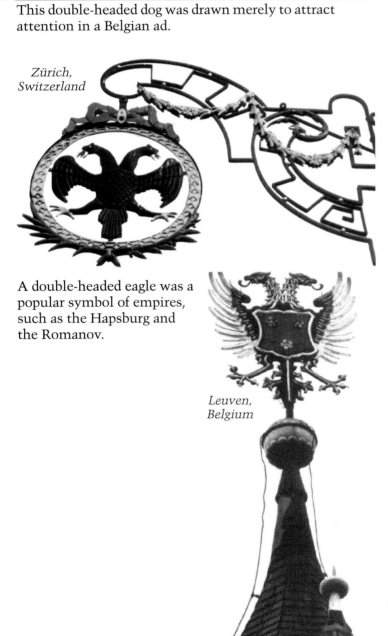

A double-headed eagle was a popular symbol of empires, such as the Hapsburg and the Romanov.

Leuven, Belgium

Prague, Czech Republic

Vienna, Austria

Toledo, Spain

George Washington's home in Mount Vernon, Virginia

In Architecture

Bilateral symmetry commonly appears in buildings of all sorts. George Washington's home in Mount Vernon, Virginia, shows approximate bilateral symmetry. However, the second window from the left, upstairs, is not really a window at all. (You discover this when you go inside the house.) Apparently the first American president was so fond of symmetry that he had the missing window painted on the outside wall, and this blind window has been preserved ever since.

Schönbrunn Palace in Vienna, Austria

The Kunjongjon Hall in Seoul, Korea

Flatiron Building in New York City, New York

11

I. BILATERAL SYMMETRY

Buckingham Palace, London, England

Heros' Square, Budapest, Hungary

BELOW: *Hungarian Parliament along the Danube River, Budapest, Hungary*

Moscow State University, Russia

The whole assembly of St. Peter's Square in Vatican City shows bilateral symmetry, which can best be appreciated when viewed from the cupola of St. Peter's Cathedral

Old gate, downtown Budapest, Hungary

Iolani Palace—the former royal residence in Honolulu, Hawaii

I. BILATERAL SYMMETRY

In Religion and Music

Religious art often embraces bilateral symmetry to express divine harmony.

ABOVE: *Venice, Italy*

RIGHT: *Church and fresco in Zagorsk, Russia*

William Blake, The Ancient of Days

Some composers rely heavily on bilateral symmetry in their works; others ignore it.

One can imagine a horizontal mirror plane in Johann Sebastian Bach's *Contrapunctus*, where the symmetry plane relates the upper and lower parts.

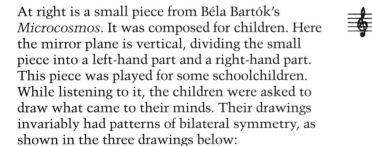

At right is a small piece from Béla Bartók's *Microcosmos*. It was composed for children. Here the mirror plane is vertical, dividing the small piece into a left-hand part and a right-hand part. This piece was played for some schoolchildren. While listening to it, the children were asked to draw what came to their minds. Their drawings invariably had patterns of bilateral symmetry, as shown in the three drawings below:

Children's drawings

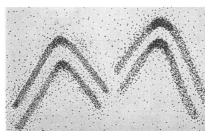

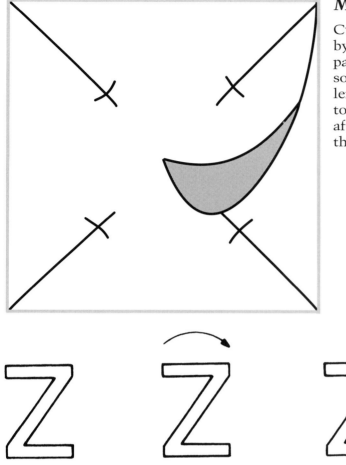

Make Your Own Pinwheel:

Cut a piece of paper into a square shape, and by folding it, find its diagonals. Then cut the paper at each corner along the diagonals to somewhat more than a quarter of their lengths. Next, fold one corner of each quarter to the center, then the other three corners one after the other, and then pin the center to a thin rod. If you blow on it, it will spin.

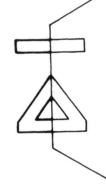

Take the letter Z, for example. It has rotational symmetry. We can rotate it around an axis, perpendicular to the paper and find that during a complete rotation, it appears in the same position twice. It has *2-fold rotational symmetry*.

On the other hand, it does not have mirror symmetry.

It is interesting to note that different kinds of symmetry may give us different feelings about motion. The presence of a symmetry plane makes things seem stationary, while rotational symmetry conveys the impression of movement.

Symmetry type
Rotational symmetry:
When an object is rotated around its axis, it appears in the same position two or more times

Symmetry operation (action)
Rotation:
The act of rotating an object around an axis

Symmetry element (tool)
Axis of rotation

Definitions
During a complete revolution, the object is reproduced:

two times	= **2-fold rotation**
three times	= **3-fold rotation**
four times	= **4-fold rotation**
five times	= **5-fold rotation**
six times	= **6-fold rotation**
etc.	

Symbols

⬬	= **2-fold rotation axis**
▲	= **3-fold rotation axis**
■	= **4-fold rotation axis**
⬟	= **5-fold rotation axis**
⬢	= **6-fold rotation axis**
etc.	

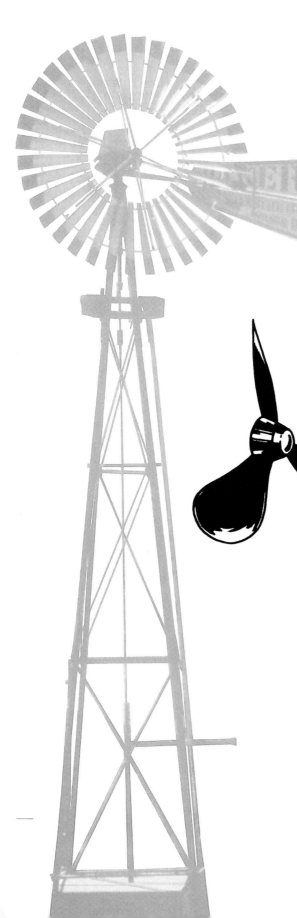

Rotating Blades

Rotating parts of various machines, such as propellers, have rotational symmetry only. The familiar windmills in Holland with four blades are very much like pinwheels. In the old days, wind power was used to rotate huge mill stones for grinding grain into flour. Rotational symmetry exclusively, as in the pinwheel or windmill, means that all the blades curve in the same direction. (This facilitates catching the wind.)

There is also a famous windmill in classical literature—the one in Spain that the self-proclaimed knight Don Quixote thought to be a four-armed giant. He tried to fight it, needless to say, unsuccessfully. (Hence the expression, "tilting at windmills.")

Recently, windmills have been utilized extensively as an alternative means of producing electricity without the need to burn nonrenewable fuels, such as oil and coal.

Two-blade propeller

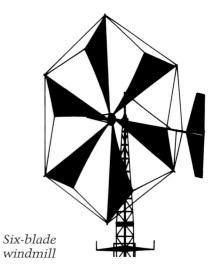

Six-blade windmill

Electricity-generating windmills, Altamont, California

Hubcap with 5-fold rotational symmetry

Wheel with 7-fold rotational symmetry

Pelton wheel of water-powered electric generator

Below are antique rotating waterwheels in front of the technical museums in two different countries. They are motionless now because they are no longer operational. By curious coincidence, both of these technical museums chose to display many-fold rotational symmetry at their entrances.

BELOW LEFT: *Rotating waterwheel in front of the Technical Museum, Oslo, Norway*

BELOW RIGHT: *Rotating waterwheel in front of the Technical Museum, Budapest, Hungary*

Two-fold, Washington, D.C.

Three-fold, Prague, Czech Republic

Four-fold, Linz, Austria

Sculptures

Rotational symmetry is also evident in these sculptures of dolphins and fish in interlocking positions. Interestingly, these occur in widely diverse parts of the world.

Taiwanese stamp

Two-fold, Rome, Italy

Logos

Bank logos often have rotational symmetry only.

Security First National Bank, 2-fold

United Bank of Colorado, 2-fold

Pittsburgh National Bank, 3-fold

American Service Bank, 3-fold

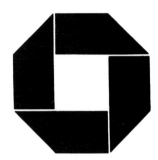

Chase Manhattan Bank, 4-fold

First American National Bank, 5-fold

Korea Housing Bank, 5-fold

Crocker Bank, 6-fold

Korea Exchange Bank, 4-fold

When our daughter Eszter was small, she suggested that banks have logos with rotational symmetry because they turn money around.

Two- and Three-Fold Rotational Symmetry

The logos of the British, Austrian, and Spanish railway systems, as well as the Tokyo and Seoul subway systems, have 2-fold rotational symmetry only, implying travel in one direction, then back.

Seoul subway

Tokyo subway

British railway logo

This is the age of the train ⇄

Vienna Airport Service ⟁ **Bahn/Rail**

Austrian railway logo

Spanish railway logo

These recycling logos have 3-fold rotational symmetry.

IL VETRO E' UNA MATERIA PRIMA

♺

RACCOLTA VETRO

INCINERATION NO! RECYCLING YES!

♺

Call Mayor Dinkins and tell him to SCC-5700 **stop garbage incineration: 212-788-3000**

The Woolmark trademark and the coat of arms from the Isle of Man have 3-fold rotational symmetry. For these logos we don't see any obvious explanation for the presence of rotational symmetry only.

Woolmark trademark

The coat of arms of the Isle of Man

Four-Fold Rotational Symmetry

Taiwanese and German stamps show 4-fold rotation

The swastika has 4-fold symmetry. It has been an ornament since prehistoric times, but it is also associated with the shameful period of Nazism and the Third Reich. It is illustrated here with a directional sign to a Buddhist temple and an anti-Nazi poster by John Heartfield.

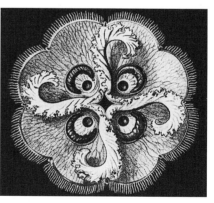

The *Aurelia insulinda* jellyfish has 4-fold rotational symmetry. This type of symmetry may be due to its circling motion in capturing food.

Sign in New York City, New York

Door handle in Jaen, Spain

Five- and Many-Fold Rotational Symmetry

NASA's symbol, shown on the wall of one of the buildings of the Florida Space Center, uses 5-fold rotational symmetry

The logo of Southern New England Telephone also has 5-fold rotational symmetry

An irregular pentagon is surrounded by tires forming a 5-fold symmetrical pattern in the advertisement of a Corpus Christi, Texas, tire shop

Logo of exchange company in Madrid, Spain, showing 8-fold rotational symmetry

This seedpod of the autograph tree (Honolulu, Hawaii) has 9-fold rotational symmetry

Manhole cover in Moscow with 16-fold rotational symmetry pattern

A United States stamp commemorating friendship with Morocco displays 12-fold rotational symmetry

Flowers

There are many flowers with only rotational symmetry. Shown here are seven such flowers from Hawaii.

LEFT: *Crown of Thorns—2-fold*

Ixora—*4-fold*

Flower of Love—5-fold

Plumeria—*5-fold*

Tiare (Gardenia Taitensis)—*6-fold*

Tiare—7-fold

Tiare—8-fold

Rotational Motifs in Buildings

Churches, synagogues, and old buildings are often decorated by motifs that have rotational symmetry only.

Milan, Italy

Small town in Italy

New York City, New York

Budapest, Hungary

Portuguese tiling, Lisbon

Italy

Folk Art

Decorations displaying exclusively rotational symmetry often occur in folk arts. Old Native American pottery has decorations with a wealth of rotational-only symmetry.

Pueblo pottery, 2-fold

Pueblo pottery, 3-fold

Mimbres pottery, 4-fold

Pueblo pottery, 4-fold

Pueblo pottery, 5-fold

Pima pottery, 5-fold

Pueblo pottery, 7-fold

Pueblo pottery, 7-fold

Mimbres pottery, 8-fold

Creating Rotational Patterns

The Spirograph toy may be used to create patterns of rotational symmetry only.

There is no symmetry plane in these patterns, only rotational symmetry, and each is chiral.

Chiral:
Describes an object that cannot be superimposed on its mirror image

The patterns can be created either left-handed or right-handed as shown here by a 5-fold motif. It is our choice which of the motifs is designated left-handed and which right-handed.

Combining Symmetries

Up to this point, we have shown two basic symmetry operations: reflection and rotation. The two of them have occurred separately, either reflection as in bilateral symmetry, or rotation as in 2-fold and many-fold rotation. For example, on page 3, the letter A had only mirror symmetry, while the letter Z on page 39 had only rotational symmetry.

The two types of symmetry elements, reflection planes and rotation axes, can also appear *together*, and in fact this often happens.

Let us look at the letter H:

Reflection:
Reflecting one-half of an object reconstructs the image of the whole object

Symmetry element (tool)
Mirror plane:
Applying a mirror plane to either of two halves, the whole is recreated

Rotation:
When an object is rotated around its axis, it appears in the same position two or more times

Symmetry element (tool)
Axis of rotation

Symbols

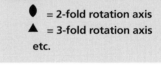

● = 2-fold rotation axis
▲ = 3-fold rotation axis
etc.

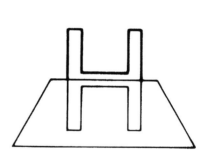
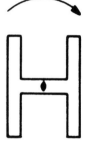

It has two reflection planes *and* one 2-fold rotation axis.

OVERLEAF: *Sculpture with 12-fold reflectional and rotational symmetry in San Francisco, California, with the Bay Bridge in the background*

Rotational and Mirror Symmetry in Flowers

Flowers show more diversity in symmetry than animals. In the animal kingdom, the most common symmetry is bilateral (or mirror) symmetry, and this implies the presence of one symmetry plane only. Most of the flowers we have looked at so far have only rotational symmetry *(see p. 47)*, but many other flowers have several **symmetry planes** as well.

This is a three-petal flower.

We can place a **mirror plane** across each petal.

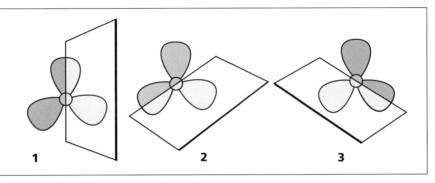

Thus, this flower has three mirror planes. These mirror planes are also called **reflection planes.**

Bilateral symmetry means the presence of one mirror plane only. When there are several mirror planes, we could call the symmetry multilateral but no one really uses this word as applied to symmetry. Generally, when we speak about mirror symmetry, it may mean the presence of one or more mirror planes, whereas *there is always only one mirror plane in bilateral symmetry.*

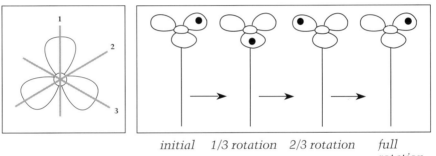

initial   1/3 rotation   2/3 rotation   full rotation (back to original)

Now let's rotate this flower around its stem. If the three petals are identical, it will look the same in its rotated position. With a three-petal flower, one-third rotation of a full revolution brings back the original flower, or at least it looks like that. The same happens with two-thirds of a full revolution. Finally, after a complete rotation, the flower not only looks the same but it is indeed back to its original position. You can mark one of the petals, at least in your imagination, to see that this is correct. Thus, this flower has 3-fold rotational symmetry.

Mirror symmetry and 3-fold rotational symmetry are present in the three-petal flower *at the same time.* In terms of symmetry, this flower is very different from the pinwheel. Whereas the pinwheel had *only* rotational symmetry, the flower has *both* reflectional and rotational symmetry.

We have already seen why the pinwheel does not have mirror symmetry, but only rotational symmetry *(see p. 38)*. It would seem natural that flowers would have both rotational symmetry and mirror symmetry. Flowers need not rotate. Thus, it is curious that some flowers, nevertheless, have rotational symmetry only and no mirror symmetry. Some botanists explain that this may be due to genetic accidents.

There is a stone carving on old ruins along Via Appia Antica in Rome with two flowers. One has only 4-fold rotational symmetry, while the other has rotational *and* mirror symmetries. Roman masons obviously used such flowers as models.

Stone carvings along Via Appia Antica, Rome, Italy

Vinca minor

Norwegian tulip

The common clover has the same symmetry as the three-petal flower. The rare four-leaf clover has four reflection planes and a 4-fold rotational axis.

Four-leaf clover on a United States stamp

Cherry blossoms in Japan

Star of Bethlehem

Carrion flower

Bougainvillea

Mussaenda

Pentas lanceolata

Yellow African iris

Lily

Daffodils

ABOVE AND BELOW: *Korean beam-end decorations with 6-fold and 5-fold symmetry*

Apple blossom with five petals

The apple core, when the apple is dissected in the plane perpendicular to its stem, shows 5-fold symmetry

The top of the cotton plant displays 5-fold symmetry

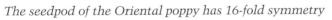

The seedpod of the Oriental poppy has 16-fold symmetry

Oriental poppy

Primitive Organisms

Primitive organisms have beautiful symmetrical shapes showing both rotational and mirror symmetry. Jellyfish seem to prefer 4-fold symmetry, as depicted here.

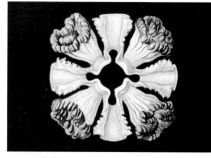

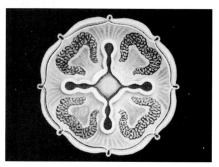

As shown above, many starfish have 5-fold symmetry.

Starfish with eleven legs

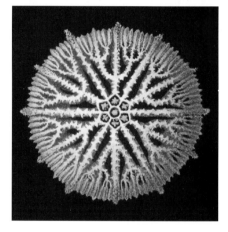

Star corals with 6-fold symmetry

LEFT: *Sea urchin with 5-fold symmetry*

RIGHT: *Sea urchin with many-fold symmetry*

Downtown Prague (early 1970s), Czech Republic

In Overhead Lighting

This streetlight has 8-fold symmetry. Besides the 8-fold rotational symmetry, there are eight symmetry planes: four going through the lamp bodies, and four going in between them.

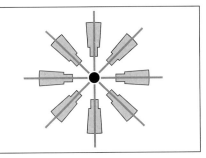
Four symmetry planes going through lamps

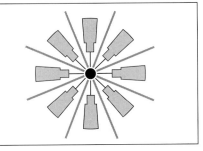
Four symmetry planes going in between lamps

This chandelier has 16-fold symmetry—again, both reflection and rotation

Streetlight in St. Peter's Square, Vatican City (6-fold symmetry—the seventh lamp is the center)

Streetlight with 4-fold symmetry, Paris, France

Logos

Logos often have both rotational and mirror symmetry.

Mitsubishi ad in Hiroshima with 3-fold symmetry, both reflection and rotation

Mitsubishi logo on a car

Logo of the Sarajevo Winter Olympics (1984), with 4-fold symmetry

The advertisement of a Sapporo, Japan, food company in the shape of a snowflake has 6-fold symmetry. (Disregard the five-pointed star in the middle that destroys the 6-fold symmetry)

The Hokkaido flag with 7-fold symmetry

Another example of Hokkaido's seven-pointed star logo, Sapporo, Japan

Ancient sculpture in Rome, Italy

In Sculpture and Display

Indian stamp

The same sculpture from India on a Soviet stamp

Three sculptures are shown here, all with 4-fold symmetry.

Hubcaps in the display of a Seoul, Korea, auto parts shop. Some hubcaps have rotational symmetry only, others have both reflection and rotation

A fountain in Pécs, Hungary

In Architecture

The cupolas of many state capitols and other important buildings have reflectional and rotational symmetry together.

State Capitol, Jackson, Mississippi

Capitol Dome, Washington, D.C.

Atomic bomb memorial, Hiroshima, Japan

St. Isaac Cathedral, St. Petersburg, Russia

Cupola of Hungarian Parliament, Budapest, Hungary

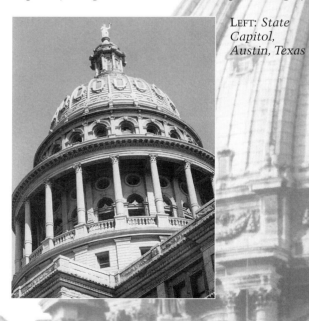

Left: *State Capitol, Austin, Texas*

St. Peter's Dome, Vatican City

V. REFLECTION & ROTATION

FAR RIGHT:
*Pantheon,
Paris, France*

NEAR RIGHT:
*One of the
two towers
of the main
synagogue in
Budapest,
Hungary*

*The Qutb
Minar in
Delhi, India*

*The world-famous leaning tower,
Pisa, Italy*

Some towers of the Vasilii Blazhennii Cathedral in Moscow show symmetries with both reflection and rotation, while others have rotational symmetry only.

V. REFLECTION & ROTATION

Radial Symmetry

When the symmetry gets so many-fold that the shape begins looking more like a circle than a many-sided polygon, we may call it **radial** or **cylindrical symmetry** *(see p. 22)*. The plant at right has "many-fold" symmetry.

Papercutting

Papercutting is a favorite children's pastime: Not only does artfully cut paper show beautiful symmetries, but symmetry makes cutting the patterns very economical. By folding the paper several times and cutting through several layers of paper, multiple identical patterns are created simultaneously.

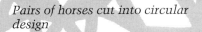

Papercutting is a traditional Chinese art. Here are two examples.

Pairs of horses cut into circular design

Pairs of monkeys cut into circular design

Fact to Consider:

Rotational symmetry, as we have seen, may appear alone, without reflection. But if an object has more than one symmetry plane, it always has rotational symmetry as well. The only case where reflection is not accompanied by rotation is when there is bilateral symmetry, with only one mirror plane.

VI. Snowflakes

*S*nowflake knew that she was beautiful. She was made up of pure, shining crystals, like fragments of glass or spun sugar. She was all stars and arrows, squares and triangles of ice and light, like a church window; she was like a flower with many shining petals; she was like lace and she was like a diamond. But best of all, she was herself and unlike any of her kind. For while there were millions of flakes, each born of the same storm, yet each was different from the other.

Paul Gallico
Snowflake

Some of the most beautiful examples of reflection and rotation in nature can be seen in snowflakes. For one to observe falling snowflakes individually, the weather must be dry and cold. When the conditions are right, the experience can be so captivating that you can't get enough of them. Snowflake after snowflake after snowflake: they are not only beautiful jewels, but each is unique unto itself.

Stamps with snowflake motifs

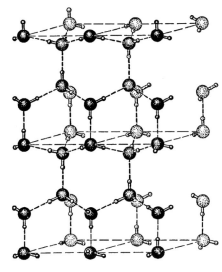

Hexagonal arrangement of water molecules in the ice crystal

Hexagonal Symmetry

The **hexagonal symmetry** of snowflakes is a consequence of the internal hexagonal structure; that is, the arrangement of the water molecules in the crystal, as shown in the drawing at left.

However, what is puzzling is that each snowflake has a different shape. Furthermore, even the smallest variations from the basic underlying shape of a snowflake are repeated in all six directions.

Each snowflake which develops unhindered has 6-fold symmetry. It has six reflection planes (three going through the branches, and three between them) and a 6-fold axis of rotation. It has other symmetries besides, one of them being a *reflection plane in the plane of the snowflake itself.* Through this plane, you could slice the snowflake into two thinner snowflakes.

Reflection:
Reflecting one-half of an object reconstructs the image of the whole object

Symmetry element (tool)
Mirror plane:
Applying a mirror plane to either of two halves, the whole is recreated

Rotation:
When an object is rotated around its axis, it appears in the same position two or more times

Symmetry element (tool)
Axis of rotation

The uniqueness of snowflakes may be related to the way they grow. Water starts crystallizing into ice in a flat 6-fold pattern and grows in six equivalent directions. As ice quickly solidifies, heat is released, and the heat flows between the branches, thereby facilitating dendritic, or treelike, growth.

Diversity in Shape

Minute differences in the local conditions of two individually growing ice crystals make them develop differently even though they may be growing next to each other. This situation is what produces the endless variety of shapes. Growing snow crystals are highly susceptible to any kind of change, and even a small force will cause spectacularly large deviations in the growing pattern of a snowflake.

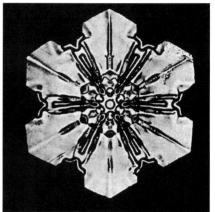

No two snowflakes are ever exactly the same

Uniform Growth

This explanation for the variations in shapes of snowflakes is a reasonable one. An amazing thing is that all minute variations occur in all six directions, something that has puzzled people for some time. Thirty years ago, an American scientist, D. McLachlan put it this way: "How does one branch of the crystal know what the other branches are doing during growth?" He noted that the kind of regularity encountered among the snowflakes is not uncommon among flowers and blossoms or among sea animals in which hormones and nerves coordinate the development of the living organisms. However, snowflakes are not living organisms, as they

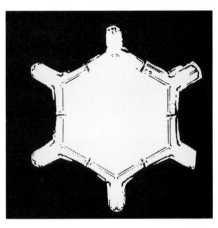

Looking Back

For ages, people have been interested in snowflakes. The oldest known recorded statement on snowflake forms dates back to the second century B.C. in China. Six was a symbolic number for water in many classical Chinese writings. The examination of snowflake shapes and their comparison with other shapes was apparently considered to be of great importance in East Asia. As a forerunner of the modern investigations into the correlation between snowflake shapes and meteorological conditions, it was noted in the thirteenth century:

The Yin embracing Yang gives hail,
the Yang embracing Yin gives sleet.
When snow gets six-pointedness,
it becomes snow crystals.
When hail gets three-pointedness,
it becomes solid.

J. Needham & Lu Gwei-Djen
Weather

McLachlan's illustration of the coordinated growth of the six branches of snowflakes based on his standing wave theory

consist of nothing other than water molecules. McLachlan's explanation for the coordination of the growth among the six branches of a snow crystal is based on thermal and acoustical waves in the crystal. These waves ensure the identical development in all six branches and this development is independent of the particular branch in which the change of the conditions occurred in the first place. The coordinated growth of the six branches produces strikingly different overall shapes but identical branches.

Johannes Kepler was the first European to recognize the hexagonal symmetry of snowflakes and in 1611 he published a small booklet entitled *The Six-cornered Snowflake*.

In 1635, René Descartes observed the shapes of snow crystals and drew them.

Among later works, William Scoresby's observations and sketches are especially important. Scoresby, who went on to become an Arctic scientist, made the drawings at right in his log book in 1806. He was 16 at the time and on a voyage with his father to the whale fisheries in Greenland.

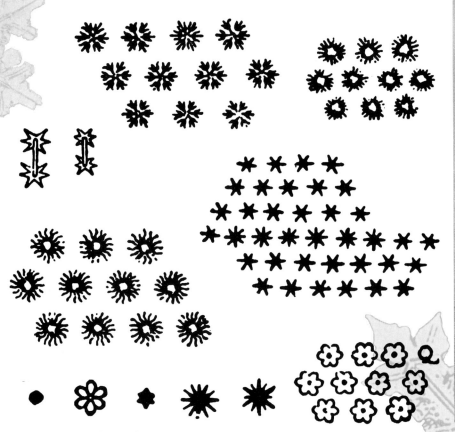

Snowflakes as drawn by René Descartes in 1635

A page from William Scoresby's log book, 1806

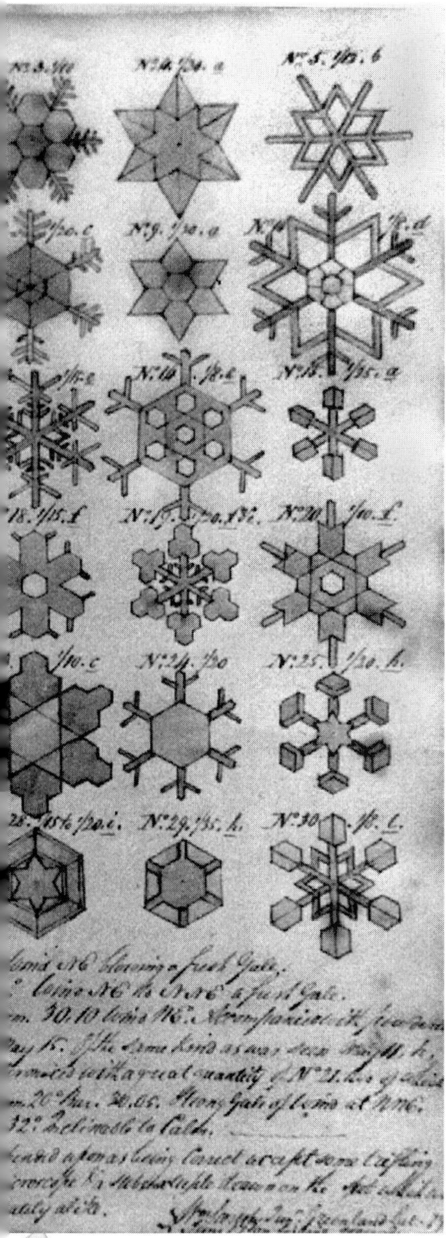

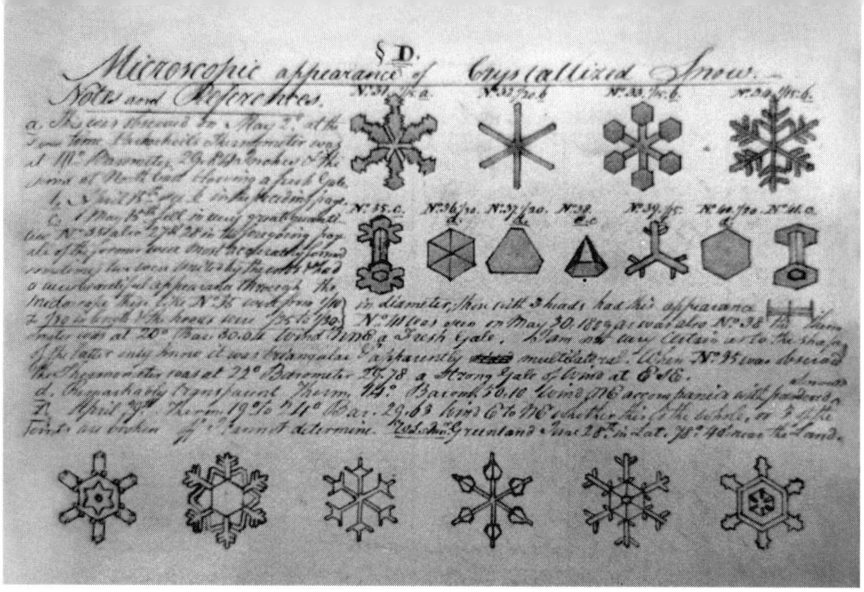
Microscopic appearance of Crystallized Snow.

Scoresby's sketches of snowflakes from his log book, 1806

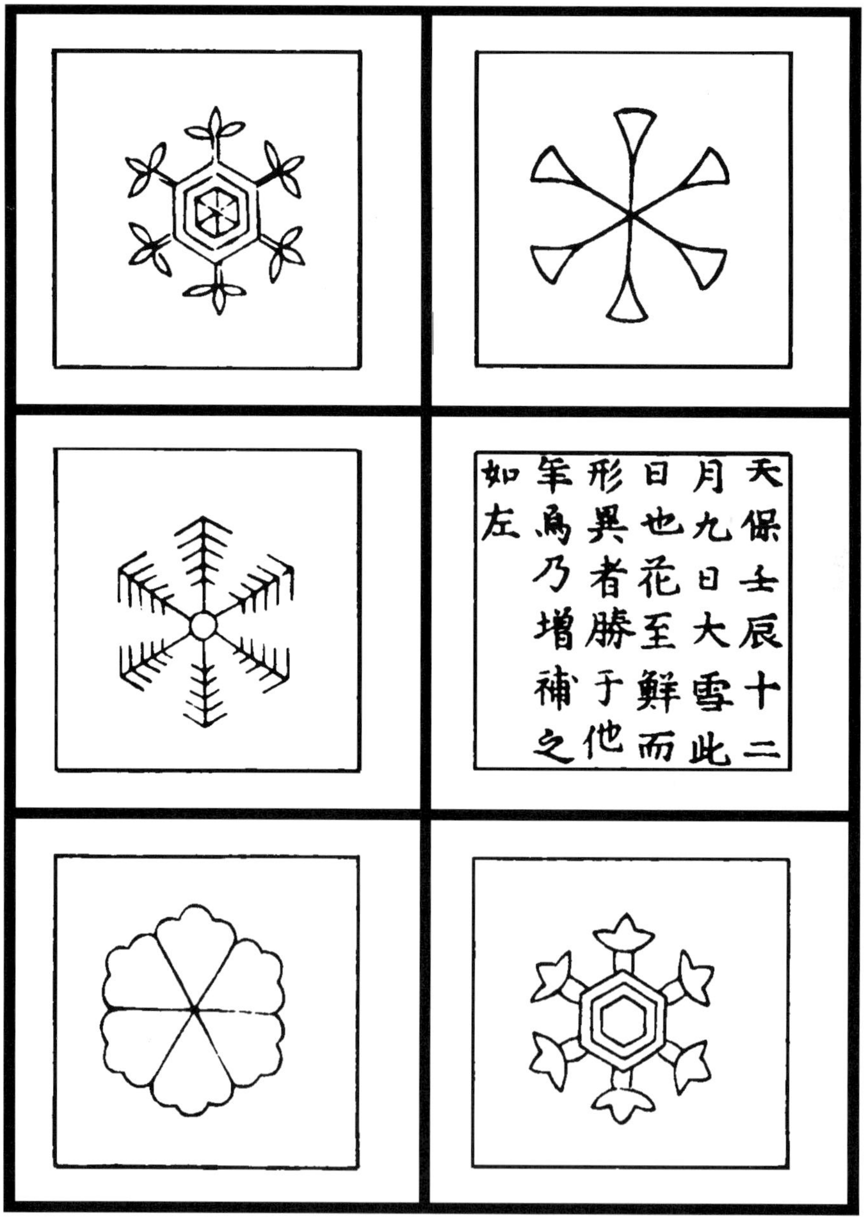
天保壬辰十二
月九日大雪此
日也花至鮮而
形異者勝于他
年爲乃増補之
如左

RIGHT: *A few years after Scoresby, in 1832, Sekka Zusetsu of Doi, Japan, did a series of excellent sketches of snowflakes*

6000 Photos

The most famous book on snow-flakes, *Snow Crystals,* by
W. A. Bentley and W. J.
Humphreys, appeared first in
1931. Bentley devoted his life to
taking photomicrographs of snow
crystals, and collected at least
6000 such photos in his work-shop at Jericho, Vermont. Over
2000 of them appeared in this
book, with text by Humphreys.
Bentley's photomicrographs have
been reproduced innumerable
times in various places, often
without credit.

W. A. Bentley photographing snow crystals

. . . the exquisite precision of form displayed by these little jewels, insignia, orders, agraffes—no jeweller, however skilled, could do finer, more minute work . . . And among these myriads of enchanting little stars, in their hidden splendour that was too small for man's naked eye to see, there was not one like unto another; an endless inventiveness governed the development and unthinkable differentiation of one and the same basic scheme, the equilateral, equiangled hexagon . . .

Thomas Mann
The Magic Mountain

Artificial Snowflakes

Another outstanding contribution is Ukichiro Nakaya's *Snow*, which first appeared in Japanese in 1938. It has been reprinted 36 times, the last time in 1987, and is now out of print. Its English version, *Snow Crystals*, was published in 1954. Working in Hokkaido, the northernmost big island of Japan, Nakaya recorded naturally occurring snow crystals, classified them, and investigated their mass, speed of fall, electrical properties, frequency of occurrence, and so on. He also developed methods of producing snowflakes artificially, and succeeded in determining the conditions of formation for different types of snowflakes.

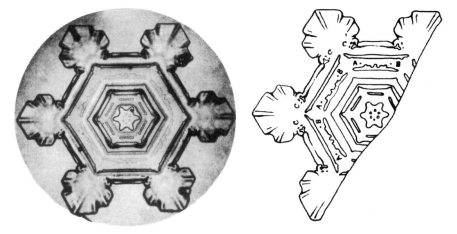

Photomicrograph of a snowflake and sketch of part of the crystal by Nakaya

RIGHT: *Nakaya's classification of snow crystals*

U. Nakaya taking measurements in his laboratory

Sculpture on the campus of Hokkaido University, Sapporo, Japan, honoring U. Nakaya and commemorating the birth of the first artificial snowflake, 1936

VII. BUILDINGS FROM ABOVE

Polygons

If we fly in an airplane over a city and look down at the buildings directly below, we see only their outlines. These shapes are **polygons**.

Regular polygons are the equilateral triangle, the square and so on; with an ever increasing number of sides, the regular polygon eventually becomes a circle.

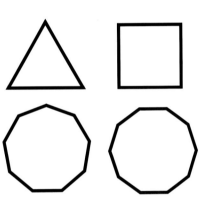

For a regular polygon, all angles are the same and all sides are of equal length. When these requirements are relaxed, polygons may appear in a great variety of irregular shapes.

OVERLEAF:
Skyscraper in Chicago, Illinois

Symmetries of Regular Polygons

First take a regular polygon with an even number of sides, for example, the regular *hexagon*. Some symmetry planes of the regular hexagon connect opposite corners, others connect the midpoints of opposite sides. Altogether, the regular hexagon has six symmetry planes this way. The intersection of these symmetry planes is in the center of the regular hexagon, and a 6-fold axis of rotation goes through this point perpendicular to the plane of the hexagon. This is, of course, the symmetry of the snowflake, among others.

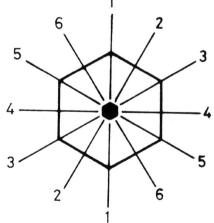

Regular polygon:
All its angles are equal and all its sides are of equal length

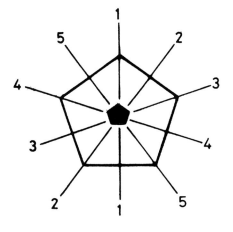

The regular *pentagon* is a regular polygon with an odd number of sides (five).

For the regular pentagon, each symmetry plane connects a corner with the midpoint of the opposite side. Altogether, it has five such symmetry planes. Then there is also an axis of 5-fold rotation going through the intersection of the symmetry planes and perpendicular to the plane of the pentagon.

Both the regular hexagon and the regular pentagon have an additional symmetry plane that is the plane of the polygon itself. In most of the previous examples, this perpendicular symmetry plane was not present. It occurs only when you see the same thing when looking from both top and bottom. Obviously this is true for the snowflake, but not for a flower or a building.

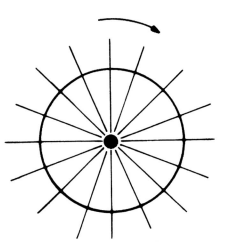

The *circle* has an infinite number of symmetry planes (only a few of them are shown on the drawing) and an infinite-fold axis of rotation:

No matter how little we turn the circle around this axis, it is enough for it to act as an axis of symmetry.

We may also say that the circle has **cylindrical symmetry** or **radial symmetry** *(see pp. 22 and 68)*. Of course, the circle has rigorous geometrical symmetry, whereas the stems of plants and the trunks of trees have only approximate symmetry.

Rotation:
When an object is rotated around its axis, it appears in the same position two or more times

Symmetry element (tool)
Axis of rotation

Symbols

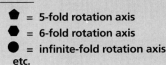 = 5-fold rotation axis

= 6-fold rotation axis

= infinite-fold rotation axis
etc.

Reflection:
Reflecting one-half of an object reconstructs the image of the whole object

Symmetry element (tool)
Mirror plane:
Applying a mirror plane to either of two halves, the whole is recreated

Famous Shapes

Now let's return to the shapes of buildings from above.

Two famous structures in Washington, D.C., have very simple, highly symmetrical outlines from above. The Washington Monument looks like a square from above with circles around it. The Pentagon *(next page)*, headquarters of the Department of Defense, expresses its shape in its name.

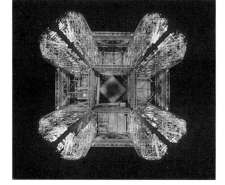

The Eiffel Tower, as seen from below at night

Washington Monument, Washington, D.C.

The Eiffel Tower in Paris, France, has a square outline

Castillo de San Marcos, St. Augustine, Florida

Goryokaku Castle, Hokkaido, Japan

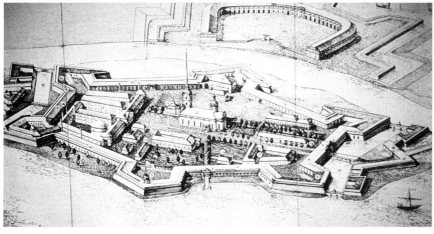

St. Petersburg, Russia

Two- to Six-Fold Symmetries

Pentagon, Washington D.C.

The Lincoln Memorial has a rectangular shape

The Lincoln Memorial is an elegant rectangle with 2-fold symmetry only, while the Pentagon and the old fortresses depicted here were also built in simple outline shapes with more symmetries.

Várad, Hungary

Castel del Monte on an Italian stamp

Eight- and Many-Fold Symmetries

A beautiful example of a polygon-shaped outline is this remarkable Italian castle—Castel del Monte in Apulia, southern Italy. It was built in the 13th century for nonmilitary purposes on the top of a hill. The outer shape is an octagon, as is the inner courtyard. Even the eight small towers have octagonal symmetry.

Castel del Monte, Apulia, southern Italy, with a multitude of regular octagonal shapes

This building, which houses a circus in Moscow, Russia, has an outline in the shape of a many-sided polygon, almost a circle

Bullfighting arena, Jaen, Spain

Jefferson Memorial, Washington, D.C.

Luzhniki Stadium, Moscow, Russia

Round-Shaped Buildings

The buildings shown here all have circular outlines for their shapes.

An ancient example of the circular outline is the Coliseum in Rome, Italy

Residential buildings in Fukuoka, Japan

Basketball court (under construction, 1989), Storrs, Connecticut

Basketball court (under construction, 1992), Fukuoka, Japan

This mosque in Acca, Israel, has both square and circular outlines

VIII. Cubes & Other Polyhedra

The Cube and Its Symmetries

In the last chapter we talked about polygons, which are two-dimensional. Now we will extend our considerations into space and discuss three-dimensional objects and their symmetries.

The cube is a three-dimensional body. It has six sides and is sometimes referred to as a hexahedron. Each of its sides is a square. The cube is highly symmetrical, because it has many different kinds of symmetry, as indicated in these drawings.

First, there are three mirror planes parallel to the sides, or faces, of the cube, as shown at right.

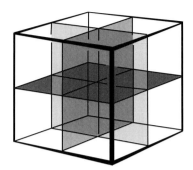

There are also other mirror planes connecting opposite edges, altogether six of them.

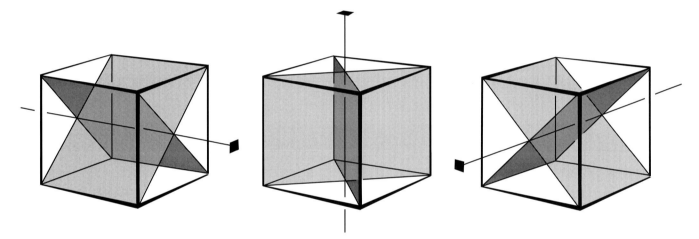

Incidentally, the crossings of these planes are themselves important symmetry elements. They are 4-fold rotational axes. They go through midpoints of opposite faces. If you rotate the cube around any one such axis, you will see the same cube four times during a complete rotation.

Reflection:
Reflecting one-half of an object reconstructs the image of the whole object

Symmetry element (tool)
Mirror plane:
Applying a mirror plane to either of two halves, the whole is recreated

Rotation:
When an object is rotated around its axis, it appears in the same position two or more times

Symmetry element (tool)
Axis of rotation

Symbols
- ⬬ = 2-fold rotation axis
- ▲ = 3-fold rotation axis
- ■ = 4-fold rotation axis
- ⬠ = 5-fold rotation axis
- ⬣ = 6-fold rotation axis

etc.

OVERLEAF: *Sculpture in New York City, New York*

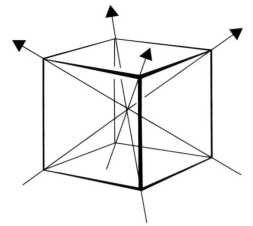

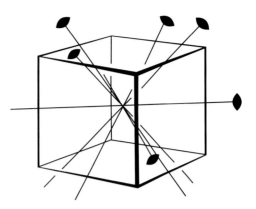

There are also 3-fold rotation axes diagonally connecting opposite vertices (corners) of the cube. There are four of them.

The 2-fold rotation axes connect midpoints of diagonally opposite edges. There are six such axes altogether.

The many symmetries of the cube may be a little hard to visualize, but if you take a cube (sugar cube, for example) and rotate it in your hands around these different axes, you will see the 2-fold, 3-fold and 4-fold symmetries. Better yet, you can make a paper model from this outline:

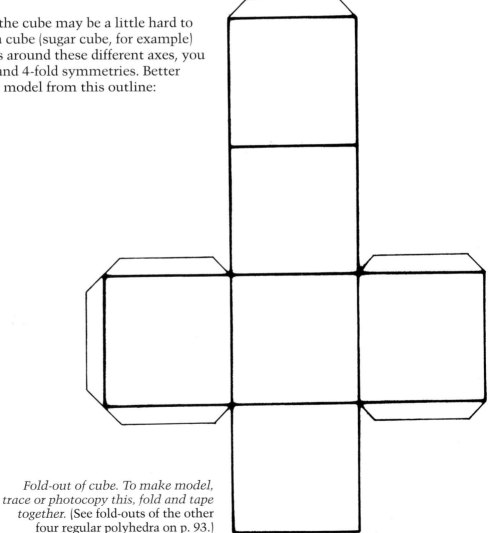

Fold-out of cube. To make model, trace or photocopy this, fold and tape together. (See fold-outs of the other four regular polyhedra on p. 93.)

The Five Regular Polyhedra

As the square is a regular polygon, the cube is a **regular polyhedron.**

We have seen that there is an infinite number of regular polygons.

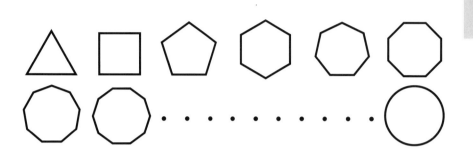

However, there are only five regular polyhedra. They are the tetrahedron, hexahedron (cube), octahedron, dodecahedron, and icosahedron. Their names indicate the number of faces:

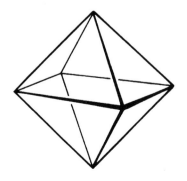

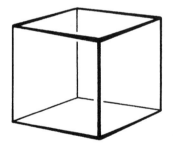

Tetrahedron = 4 faces

Hexahedron (cube) = 6 faces

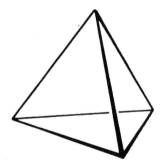

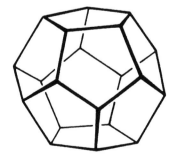

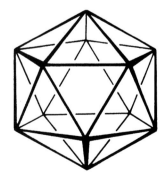

Octahedron = 8 faces

Dodecahedron = 12 faces

Icosahedron = 20 faces

These regular polyhedra constituted an important part of Plato's natural philosophy and are thus also called **Platonic solids** or Platonic bodies. (Plato lived from 427–347 B.C.)

The tetrahedron, cube, and octahedron are relatively simple shapes, while the discovery of the dodecahedron and the icosahedron was referred to as ". . . one of the most beautiful and singular discoveries made in the whole history of mathematics." (Hermann Weyl, *Symmetry*, 1952)

H. S. M. Coxeter visiting at Smith College, Northampton, Massachusetts, 1984

The regular polyhedra have been known from time immemorial. In fact, H. S. M. Coxeter, Professor of Mathematics at the University of Toronto, who was called "the geometer of the 20th century," likened the question of who first constructed the regular polyhedra to asking the question of who first used fire.

. . . the chief reason for studying regular polyhedra is still the same as in the times of the Pythagoreans, namely, that their symmetrical shapes appeal to one's artistic sense.

H. S. M. Coxeter
Regular Polytopes, 3rd ed.
1973

Characteristics of the Regular Polyhedra

Name	Shape of Faces	Number of Faces	Number of Vertices	Number of Edges
Tetrahedron	Triangle	4	4	6
Cube	Square	6	8	12
Octahedron	Triangle	8	6	12
Dodecahedron	Pentagon	12	20	30
Icosahedron	Triangle	20	12	30

Euler's formula: $v + f = e + 2$
For example, the octahedron:
v (vertices) $+ f$ (faces) $= e$ (edges) $+ 2$;
$6 + 8 = 12 + 2$

Duality

The cube and the octahedron are dual to each other. Both have 12 edges. The cube has 6 faces and 8 vertices, and the octahedron has 8 faces and 6 vertices.

The dodecahedron and icosahedron are also dual to each other. Both have 30 edges; the dodecahedron has 12 faces and 20 vertices, while the icosahedron has 20 faces and 12 vertices.

The tetrahedron stands alone among the five. (It has no dual polyhedron.)

Symmetries of the Regular Polyhedra

Only a few characteristic symmetry elements are shown for each of the five regular polyhedra. For example, with the cube, we are showing only one each of the 2-fold, 3-fold, and 4-fold rotation axes and only two of the mirror planes.

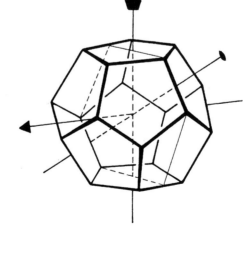

Models of the Regular Polyhedra *(Opposite)*

Again, the symmetries of the regular polyhedra are easier to see if we take them into our hands. On the opposite page are patterns that you can use to construct four of the five regular polyhedra. *(See p. 89 for the cube pattern.)* You can make these out of paper by tracing, or you can use cardboard for sturdier models.

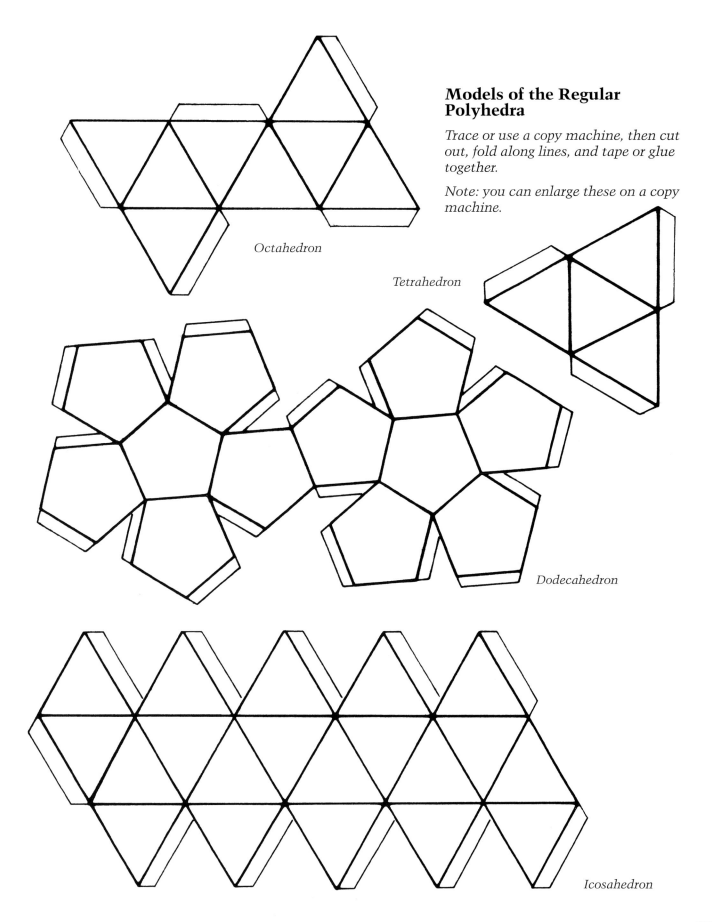

Models of the Regular Polyhedra

Trace or use a copy machine, then cut out, fold along lines, and tape or glue together.

Note: you can enlarge these on a copy machine.

Octahedron

Tetrahedron

Dodecahedron

Icosahedron

Regular Polyhedra in Nature

The cube is the best known among the regular polyhedra in the world around us.

Many primitive organisms are shaped like regular polyhedra, such as radiolarians.

Radiolarians with polyhedral shapes

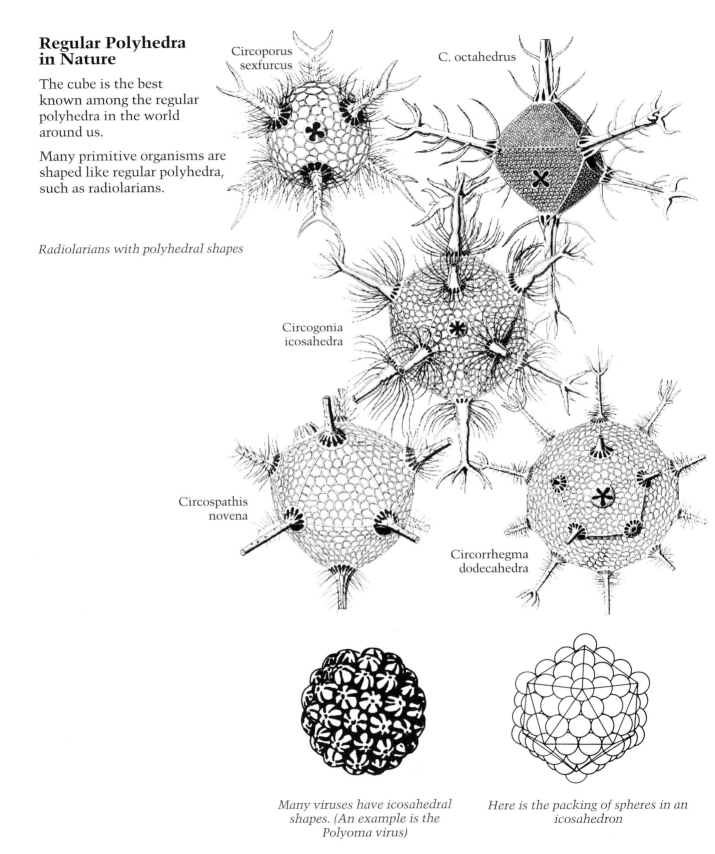

Circoporus sexfurcus

C. octahedrus

Circogonia icosahedra

Circospathis novena

Circorrhegma dodecahedra

Many viruses have icosahedral shapes. (An example is the Polyoma virus)

Here is the packing of spheres in an icosahedron

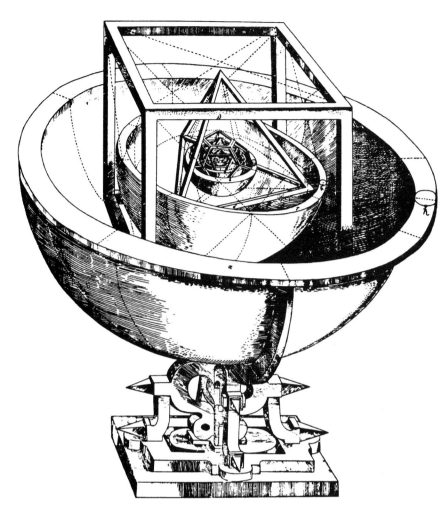

The Regular Polyhedra in a Planetary Model

Regular polyhedra have always fascinated people. Sometimes people would infer their presence even when they were not really there. For example, in the 16th century, when planetary motion was not yet understood, Johannes Kepler prepared a model in which the regular polyhedra were nested within each other.

According to Kepler's planetary model, the greatest distance of one planet from the sun stands in a fixed ratio to the least distance of the next outer planet from the sun. Only six planets were known in Kepler's time and he described their distances by five such ratios. A regular solid can be interposed between two adjacent planets so that the inner planet, when at its greatest distance from the sun, lies on the inscribed sphere of the solid, while the outer planet, when at its least distance, lies on the circumscribed sphere.

Today, of course, we know that this model is wrong. However, it is a beautiful model and symbolizes Kepler's attempt at attaining a unified approach to such diverse branches of the sciences as (what we call today) astronomy and crystallography.

Kepler's planetary model, in which all five regular polyhedra are used to describe the trajectories (paths) of the six planets known at that time. (Johannes Kepler, Mysterium Cosmographicum, *1595)*

Kepler's Ratios			
	Ratio of Inscribed to Circumscribed Sphere (x 1000)	Ratio of Inner to Outer Planetary Orbit (x 1000) Using the Copernican Distances	
		1000	Saturn
Cube	577	572	Jupiter/Saturn
Tetrahedron	333	290	Mars/Jupiter
Dodecahedron	795	658	Earth/Mars
Icosahedron	795	719	Venus/Earth
Octahedron	577	500	Mercury/Venus

In *Harmonices Mundi,* Kepler used the five regular polyhedra to represent what people considered in his time the four elements and the universe:

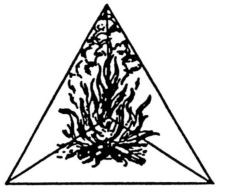

Tetrahedron
Fire

The five regular solids drawn by Johannes Kepler in Harmonices Mundi, Book II, *1619*

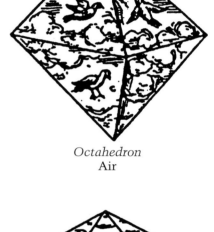

Octahedron
Air

Cube
Earth

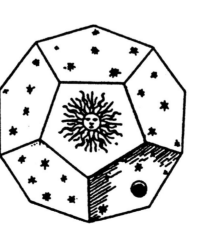

Icosahedron
Water

Dodecahedron
The Universe

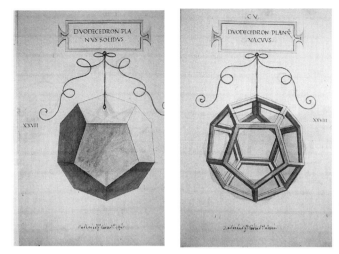

Leonardo da Vinci, illustrations to Luca Pacioli: De Divina Proportione, *1509*

Artistic Dodecahedra and Icosahedra

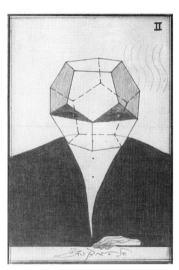

Salvador Dali holding a pentagonal dodecahedron. Drawn by Ferenc Lantos after a photograph

Artists have often attributed mysterious qualities to the pentagonal dodecahedron.

Horst Janssen, Crystal-Slave

Herbert Hauptman, (Chemistry Nobel Laureate, 1985) and two of his stained glass models. ABOVE LEFT, *icosahedron.* AT RIGHT, *pentagonal dodecahedron*

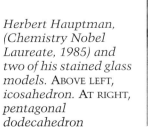

Star Polyhedra

The five regular polyhedra are convex polyhedra. Convex means having surfaces that bulge outward. Thus, the angles formed by any two faces joined along a common edge are always less than 180 degrees. If we remove this restriction, there are four more regular polyhedra, called regular **star polyhedra.**

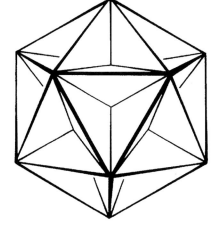

Great dodecahedron

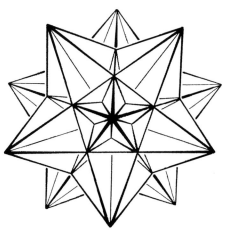

Great icosahedron

Small stellated dodecahedron

Great stellated dodecahedron

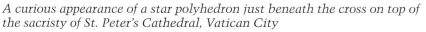

A curious appearance of a star polyhedron just beneath the cross on top of the sacristy of St. Peter's Cathedral, Vatican City

Archimedean Polyhedra

In addition to the regular poly-hedra, there are various families of polyhedra with decreased degrees of regularity. One such family is the thirteen so-called **semiregular polyhedra**, shown here. It is believed that they were first described by Archi-medes; therefore they are also called **Archimedean polyhedra.**

Characteristics of the Archimedean polyhedra:

- All their faces are regular polygons

- Their vertices are all alike

- Their faces are not all of the same kind. (This is where they differ from the regular polyhedra)

The simplest semiregular polyhedra are obtained by symmetrically shaving off the corners of the regular solids:

Truncated tetrahedron

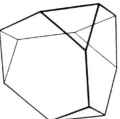

Truncated cube

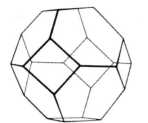

Truncated octahedron

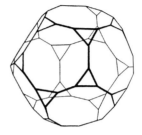

Truncated dodecahedron

Truncated icosahedron

The other eight Archi-medean polyhedra are shown here:

Cuboctahedron

Rhombicuboctahedron

Greater rhombicuboctahedron

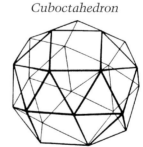

Snub cuboctahedron

Icosidodecahedron

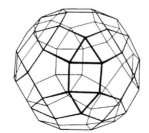

Rhombicosidodecahedron

Greater rhombicosi-dodecahedron

Snub icosidodecahedron

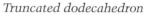

The Buckyball Molecule

Today, the truncated icosahedron is an exceptionally important polyhedron due to the recently discovered C_{60} molecule called *buckminsterfullerene*, or in short, buckyball. Many names were proposed for this newly discovered substance. One of them in Europe was "footballene." However, what Europeans call football, Americans call soccer, so in the United States this would have to be translated as "soccerene."

The molecule was named after R. Buckminster (Bucky) Fuller, the inventor and designer who used icosahedral geometry as the basis for his geodesic domes.

ABOVE: *R. Buckminster Fuller's geodesic dome at the Montreal Expo in 1967*

LEFT: *R. Buckminster Fuller (1895–1983) at Pacific High School, Saratoga, California, 1970*

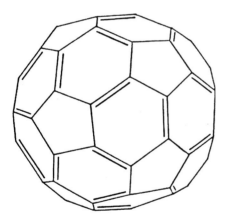

The structure of the superstable C_{60} molecule in which the tetravalency of all carbon atoms is neatly maintained

Ivory Coast stamp honoring the football world championship, Argentina, 1978

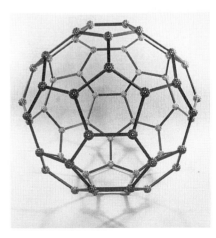

ABOVE: *Truncated icosahedron model made with Steve Baer's Zometool model kit*

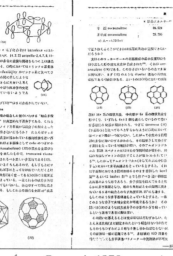

A page from Osawa's 1970 paper in the Japanese journal Kagaku

A page from Bochvar and Gal'pern's 1973 paper in the Russian journal Dokl Akad. Nauk SSSR

Fuller's ideas inspired the chemists who discovered this substance in 1985 to theorize that its structure was a truncated icosahedron. In 1991, this hypothesis was proven valid. An interesting footnote in science history, as it turns out, was that a Japanese scientist, E. Osawa, in 1970, and two Russian scientists, Bochvar and Gal'pern, in 1973, conjectured on the possibility of such a molecule on the basis of symmetry considerations. Alas, their papers were published only in Japanese and Russian.

Buckminsterfullerene was named "Molecule of the Year" in the December, 1991, issue of *Science* magazine and made the cover of numerous other journals as well, of which only a small sample is presented here.

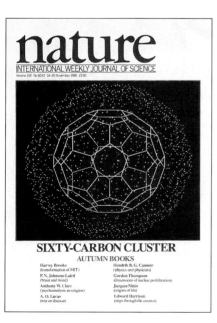

Prisms and Antiprisms

Another family of polyhedra is
the prisms and antiprisms, and
there is an infinite number of
them. A prism has two equal
and parallel faces that are joined
by parallelograms. An antiprism
also has two equal and parallel
faces, but they are joined by
triangles.

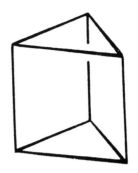
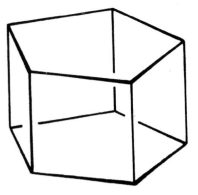
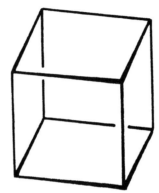
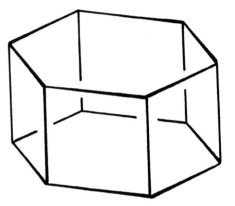

Prisms

Antiprisms

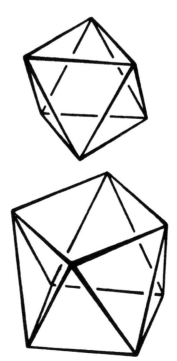
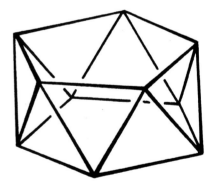
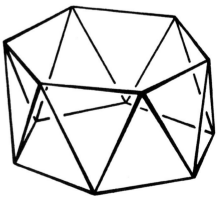

Earth as photographed from the Apollo 17 spacecraft during the final lunar landing mission in NASA's Apollo program. This view extends from the Mediterranean Sea area to the Antarctica south polar ice cap. Note the heavy cloud cover in the Southern Hemisphere. Almost the entire coastline of the continent of Africa is clearly delineated. The Arabian Peninsula can be seen at the northeastern edge of Africa. The large island off the southeastern coast of Africa is the Malagasy Republic. The Asian mainland is on the horizon toward the northeast

Statue in front of the World Trade Center, New York City, New York

The Sphere

Finally, the sphere deserves mention. It is one of the simplest possible figures, which is why it has an unlimited amount of symmetry. For example, any of its diagonals is an infinite-fold rotational axis and there is an infinite number of such diagonals. It also has an infinite number of reflection planes going through any of the diagonals. We have already seen examples of spherical symmetry from nature on page 23 (although those examples illustrate not geometric, but approximate spherical symmetry).

Gas storage tank in Inchon, Korea

. . . the spherical is the form of all forms most perfect, having need of no articulation; and the spherical is the form of greatest volumetric capacity, best able to contain and circumscribe all else; and all the separated parts of the world—I mean the sun, the moon, and the stars—are observed to have spherical form; and all things tend to limit themselves under this form—as appears in drops of water and other liquids—whenever of themselves they tend to limit themselves. So no one may doubt that the spherical is the form of the world, the divine body.

Copernicus,
De Revolutionibus Orbium Caelestium, 1543

Polyhedra in Sculptures

Simple though they may be, polyhedral shapes are frequently used in modern sculpture with intriguing results.

RIGHT: *New York City, New York*

FAR RIGHT: *Sapporo, Japan*

Milan, Italy

Inha University Campus, Inchon, Korea

Sculpture in Pécs, Hungary, by Victor Vasarely

A wall painting in New York City, New York, in the early '80s

The Pyramid at the Louvre, Paris, France

Statue of Constitution, Madrid, Spain

Albany, New York

Garden lantern in the Shugakuin Imperial Villa, Kyoto, Japan, with cuboctahedron top decoration

Modern street lanterns in Sapporo, Japan

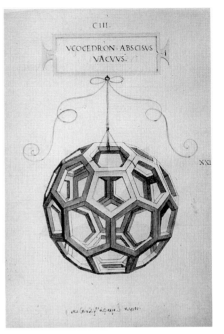

Leonardo da Vinci, illustration to Luca Pacioli: De Divina Proportione, 1509

Decoration of a marketplace in Moscow, Russia

Playground in Tel Aviv, Israel

Truncated icosahedron climber in a playground on the campus of Hokkaido University, Sapporo, Japan

Balloons

We described various polyhedral shapes in the last chapter. Now we are going to see how clusters of various objects in their actual settings form different polyhedral shapes. We shall also investigate why these polyhedral shapes appear so frequently.

We'll start by asking what happens when things have to arrange themselves on their own within a limited volume. What are the shapes they form? What are the symmetries?

We can tell a lot by connecting balloons in small groups. Balloons are very flexible, commonly available, and work well in demonstrating certain geometrical principles. Balloons come in various shapes; some are long and rather narrow, like hot dogs, others are more or less round. We'll use the round type here; as you'll see, this shape means the balloons have to elbow each other for space when they're connected.

Let's see what happens when we form small groups of two, three, four, five, and six balloons. We'll connect the balloons at their navels.

Two balloons lie along a straight line when connected.

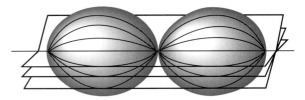

We can imagine as many symmetry planes through them as we like.

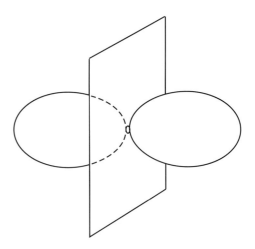

There is also a symmetry plane reflecting the two balloons into each other.

Two balloons make a straight line

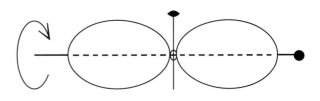

Then there is an infinite-fold axis of rotation along their connecting line, and an infinite number of 2-fold rotation axes going through the connecting point. They are perpendicular to the infinite-fold rotation axis (only one of these 2-fold axes is shown).

OVERLEAF: *Cluster of four walnuts growing together*

 Three balloons make an equilateral triangle

 Four balloons make a tetrahedron

Three balloons form an equilateral triangle. This triangle has all the symmetries that the three-petal flower has *(see p. 53)*, and even more: it also has one symmetry plane that is perpendicular to the other three planes and bisects the three balloons.

Four balloons connected together take the shape of a tetrahedron, one of the five regular polyhedra.

Five balloons are a little difficult to connect. To do this, first form a group of two balloons and then another of three, then put the two groups together.

The overall shape is of two triangle-based pyramids joined at their base. It is called a *trigonal bipyramid*.

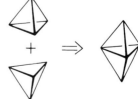

109

Six balloons can be connected by bringing a group of two and a group of four together. This gives us an octahedron, another one of the five regular polyhedra.

Incidentally, the octahedron is also a *tetragonal bipyramid*; that is, two square-based pyramids joined at their base.

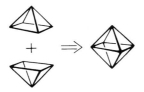

To summarize, when the balloons are connected at the navels, they will naturally cluster in these shapes:

Number of Balloons	Arrangement
Two	A line
Three	Equilateral triangle
Four	Tetrahedron
Five	Trigonal bipyramid
Six	Octahedron

Since these connected groups of balloons are flexible, we can force them to form other shapes. For example, the tetrahedral shape of the four balloons can be forced into a square planar shape. However, as soon as we stop interfering with the cluster's natural tendencies and leave the balloons alone, they immediately rearrange themselves into the previous tetrahedral shape.

The Origin of Shapes

When these balloons cluster, they are rather crowded around their connecting point. They seem to be elbowing each other for space until they assume the most economical positions about the connecting point.

This whole question can be reduced to the simple mathematical problem of arranging points on the surface of a sphere in such a way that the points are at maximum distances from each other.

Why is this? Each of the components, here balloons, takes up space. So the best arrangement will be when they are as far from each other as possible (allowing each maximum space). The arrangements of up to six points, shown for the balloons on the previous pages and shown for the points on a sphere at right, are the best for utilizing available space with maximum efficiency. However, for more than six points, or more than six balloons, there may be several arrangements about equally good for utilization of available space, so it is difficult to predict what arrangement or shape to expect.

Two points are at the two ends of a diagonal of the sphere, forming a line

Three form an equilateral triangle along a circumference of the sphere

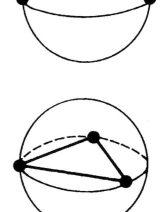

Four points will have the shape of a tetrahedron

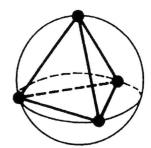

Five form a trigonal bipyramid

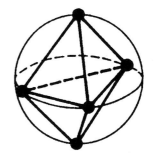

Six make an octahedron

Walnuts

Sometimes walnuts grow together in small clusters on trees. Most of them grow singly or in twos, but threes and even fours are not uncommon. A cluster of five is unusual, and six is truly rare. Groups of chestnuts exhibit similar patterns, but we have yet to see larger clusters.

Walnuts

It is not surprising that the walnut clusters have the same shapes as the balloon groups. Just as the balloons group themselves (due to their elasticity), so do the walnuts (as they slowly grow), elbowing each other for space to find the most advantageous arrangements. These arrangements are the ones where they best utilize the available space, as the points-on-the-sphere model has shown on the previous page.

This is but one example showing that the forms and shapes in nature develop according to some underlying principles, among which the *need for space* is of primary importance.

Chestnuts

Molecules

The previous considerations on shape find an important application in chemistry in understanding the structure of molecules. A **molecule** is the smallest part of a substance that can exist in a free state and still retain its chemical identity (such as how it reacts with other substances). Molecules consist of atoms held together by strong bonds. For example, the water molecule consists of three atoms—one oxygen and two hydrogens. When they say that a new drug has been synthesized, it means that molecules of a new substance were made. Molecules may react with each other to produce new molecules.

We already mentioned molecules in chapter III *(p. 32)* where we discussed their handedness (chirality). We also mentioned that molecular chirality was important in determining the behavior of molecules. The chirality of molecules is part of the spatial structure describing the direction of arrangement of their constituent atoms.

Generally, it is important to know the shape of molecules. It is more than just the order in which the atoms are arranged, it is also the shape of this arrangement. This three-dimensional arrangement of the atoms in a molecule is the structure of the molecule. All properties of a molecule are closely related to its structure.

The atoms in the molecule are linked together by pairs of electrons. Usually each of the two atoms being linked together contributes an electron to this linkage. A pair of electrons can make the two atoms stick to each other very strongly.

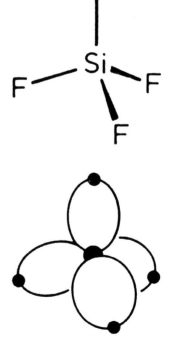

Usually the electron pair linking two atoms is depicted as a straight line between the symbols of the two atoms in a molecule. Thus, for example, in the silicon tetrafluoride molecule, there is a silicon atom in the middle and there are four linkages connecting each of the four fluorine atoms to the silicon atom. Each of these linkages is symbolized as Si–F.

Silicon tetrafluoride molecule

This is convenient but not very realistic, because the electron pair connecting the silicon and fluorine atoms is better expressed as a certain domain in space, rather than a thin line. This bond of the electron pair takes up some space that might be compared to a balloon or a walnut, rather than to a thin line. Having grasped this concept, we are then not surprised to find that the arrangement of the *four* electron pairs (that is, of the four Si–F bonds about the silicon atom) will be tetrahedral.

A more realistic version of the above molecule, showing the four linkages in space

Thus, by analogy:

The *two* bonds in magnesium difluoride will be along a straight line.

The *three* bonds in aluminum trifluoride form an equilateral triangle.

The *five* bonds in phosphorus pentafluoride make a trigonal bipyramid.

The *six* bonds in sulfur hexafluoride form an octahedron.

(*See the balloon groupings, pp. 108 – 110.*)

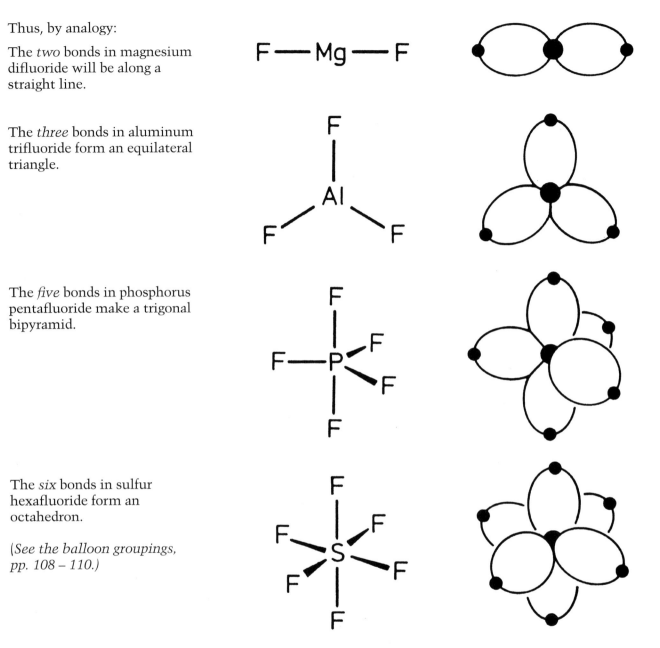

While we could take the balloons and the walnuts into our hands, molecules are fantastically small. They are not visible under a microscope with 100 or even 1 million times magnification. (A molecule is about as much smaller than the head of a pin as the head of a pin is smaller than the whole earth.) Yet scientific experiments and computations allow scientists to determine the shapes of molecules and measure their sizes.

The origin of form and shape and their symmetries make a natural connection among the balloons, walnuts, and molecules. Groups of balloons and walnuts do not come in many varieties, but molecules do. So it is very convenient to establish some simple and generally valid rules on the basis of our observations. These rules then help us predict the shape of molecules without the necessity of carrying out complicated experiments and computations each time we want to determine these shapes.

X. ANTISYMMETRY

X. ANTISYMMETRY

The symmetry of opposites is called **antisymmetry**. Each symmetry has its corresponding antisymmetry. Antisymmetry means that a property (color, for example) turns into its opposite during the symmetry operation, as when we apply a mirror.

Here a swan looks into a mirror and we see its reflection.

Suppose the same swan looks into an imaginary mirror—one that will not only reflect the swan's image, but will also reverse the black and white colors. Let us call this imaginary mirror an **antimirror** and this process **antireflection.**

Antisymmetry:
Symmetry of opposites

Reflection:
Reflecting one-half of an object reconstructs the image of the whole object

Symmetry element (tool)

Mirror plane:
Applying a mirror plane to either of two halves, the whole is recreated

Antireflection:
Reflection accompanied by property reversal

Symmetry element (tool)
Antimirror

116 SYMMETRY

Antimirror Symmetry

Here are a few more illustrations of antimirror symmetry:

1 & 2   are related by mirror symmetry
3 & 4   are related by mirror symmetry
1 & 4   are related by antimirror symmetry
2 & 3   are related by antimirror symmetry

Hungarian wine ad

Positive and negative pictures of an Eastern Orthodox church in Zagorsk, Russia

117

Antirotational Symmetry

Color changes may be introduced by rotational symmetries as well. Antirotational symmetry means that, during rotation, the object reoccurs more than once, but with one of its properties reversed at each step.

Thus, here we have 2-fold, 4-fold, and 6-fold antirotational symmetry.

2   4   6

Antisymmetry in the Universe

A natural symmetry of opposites seems to be built into the very fabric of existence. Modern physics has discovered the presence of antimatter. When matter as we know it collides with its antimatter counterpart (an electron with an antielectron, say), the two are annihilated. The universe is still here only because, as some theories hold, antimatter is in extremely short supply. Most of it did not even survive the first microsecond of the Big Bang, which supposedly brought the cosmos into being.

Interestingly, the Chinese Taoists had a similar theory about creation. They believed a unified cosmic force split into two opposing parts called Yin and Yang. These represent naturally occurring dynamic energies that are in opposition: night/day, hot/cold, male/female, young/old, etc. The symbol for Yin/Yang looks like this:

Notice that a small area of black swirls to a large area that contains a dot of white, and vice versa. This means that black and white (or whatever energies they represent) carry within them the seed of their opposite, so that when, in their movement, they reach the extreme limit, they turn into their opposite.

The central motif in the flag of the Republic of Korea is an example of the same 2-fold anti-rotational symbol. It is thought that the wavy shape of division between the two halves (resulting in rotational symmetry, rather than reflection) conveys the feeling of harmony between the two halves and not merely the contrast between them.

Rotation:
When an object is rotated around its axis, it appears in the same position two or more times

Symmetry element (tool)
Axis of rotation

Antirotation:
Rotation accompanied by property reversal at each step

Symmetry element (tool)
Antirotation axis

More Subtle Examples of Antisymmetry

So far we have seen examples with geometrically rigorous rules of antisymmetry. Even in the last example, the Yin/Yang motif is geometrically rigorous. However, when we consider other properties of this symbol, for example, male/female, we depart from strict geometrical rules: here we have the symmetry of opposites, but they are not antisymmetrical in the same rigorous sense as is a color reversal.

In the next example, the two ballet dancers show 2-fold antirotational symmetry in which there is not only color change, but gender change as well. Here we may consider either color or gender as the property being reversed during rotation. Please note, however, that there is no strict geometrical relationship between the two "parts," that is, the female and male dancers. Yet we have no difficulty in perceiving an antisymmetric relationship here.

Russian ballet dancers

"This is perestroika to some." (Soviet poster, 1987)

There was an award-winning poster in Moscow in 1987 entitled "This is perestroika to some." *Perestroika* is the Russian word for "restructuring," introduced by then–President Mikhail Gorbachev in the mid-'80s, when he thought that structural reforms could save his country. This poster apparently implied dissatisfaction with the way it was being carried out. The designer of the poster used a simple color reversal to illustrate the lack of substantial changes.

As we have seen with the ballet dancers, it is not only colors that may change. Antisymmetry may involve any kind of property. It is important only that we specify what property is changing into its opposite. For example, both in the Vasarely drawing below and in the decoration of this car, antireflection causes not only the colors to reverse but the circles to change into squares, and vice versa as well.

Vasarely painting on a French stamp

Print by Victor Vasarely

Op-art style paint job

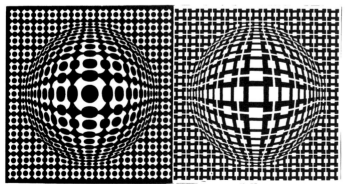

The antireflection principle changes winter into summer in the logo of a Boston sporting goods store. (Part of the logo is half of a snowflake, the other part is half of the sun.)

Shop logo in Boston, Massachusetts

Here is antisymmetry in a gas station. Not only do the colors of the letters reverse, but a more important property is reversed: the type of service provided.

Gas station in Oahu, Hawaii

You could think of these two Coke machines as being related by antimirror symmetry with a color reversal. And, actually, it's not only the colors that change into opposites. There is another, more important property being reversed here: the sugar content of regular Coke and diet Coke.

Coke machines related by color reversal and sugar content reversal

Although the examples on this page all display the symmetry of opposites, there is again no strict geometrical correspondence. When an antimirror plane is positioned between the two halves of the Herman's logo, or between the man and woman in the Russian sculpture, there is no geometrical correspondence between the two parts.

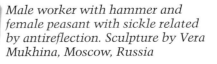

Male worker with hammer and female peasant with sickle related by antireflection. Sculpture by Vera Mukhina, Moscow, Russia

Antisymmetry in Geography

Most of our readers are probably from the Northern Hemisphere, so James Reston's description of New Zealand will be easily perceived as an expression of antisymmetry:

. . . Nothing is quite the same here. Summer is from December to March. It is warmer in the North Island and colder in the South Island. The people drive on the left rather than on the right. Even the sky is different—dark blue velvet with stars of the Southern Cross—and the fish love hooks . . .

James Reston
International Herald Tribune, 1981

Antisymmetry as a Literary Device

Some of the greatest writers have employed the symmetry of opposites as a narrative technique to evoke a mood or to describe a situation. Look at the oft-quoted opening of *A Tale of Two Cities* by Charles Dickens:

It was the best of times, it was the worst of times, it was the age of wisdom, it was the age of foolishness, it was the epoch of belief, it was the epoch of incredulity, it was the season of Light, it was the season of Darkness, it was the spring of hope, it was the winter of despair, we had everything before us, we had nothing before us, we were all going direct to Heaven, we were all going direct the other way —in short, the period was so far like the present period, that some of its noisiest authorities insisted on its being received, for good or evil, in the superlative degree of comparison only.

Or take the following example from the short story "Two Diagnoses" by Frigyes Karinthy, a Hungarian writer of the 1930s. The same person goes to see a physician at two different places on two different occasions. At the recruiting station he would obviously like to avoid getting drafted, while at the insurance company he would like to acquire the best possible terms for his policy. His answers to identical questions of the physicians are related by antisymmetry. (This is an edited excerpt.)

	At the recruiting station	**At the insurance company**
	Broken-looking, sad, ruined human wreckage, feeble masculinity, haggard eyes, shaky movement.	*Young athlete with straightened back, flashing eyes, firm movement.*
How old are you?	Old . . . very old, indeed.	*(Coyly)* You know, I'm almost ashamed to be so young . . .
Your I.D. says you're 32.	*(With pain)* To be old is not to be far from the cradle—but near the coffin.	To be young is not to be near the cradle—but far from the coffin.
Are you ever dizzy?	Don't mention dizziness, please, Doctor, or I'll collapse at once. I always have to walk in the middle of the street, because if I look down from the curb, I become dizzy at once.	Quite often, sorry to say. Every time I'm aboard an airplane and it's upside-down, and breaking to pieces. Otherwise, not . . .

X. ANTISYMMETRY

And finally, use your imagination in the following anecdote. Can you see an antisymmetrical connection?

A Marquis at the court of Louis XIV enters his wife's boudoir and finds her in the arms of a Bishop. The Marquis then walks calmly to the window and goes through the motions of blessing the people in the street.

"What are you doing?" cries his anguished wife.

"Monseigneur is performing my functions," replies the Marquis, "so I am performing his."

Arthur Koestler
The Act of Creation, 1964

Angel and devil, Notre Dame Cathedral, Paris, France

XI. REPEATING EVERYTHING

Translational Symmetry

In border decorations, a pattern can be generated simply by repeating a motif. This is symmetry again, but a very different kind of symmetry from what we have seen so far.

The symmetry operation here is **translation**. The principle is the same as in reflection or rotation. We have a simple means of creating the same thing again when in a different position. **Translational symmetry** means shifting and repeating the motif —the resulting pattern is periodic. **Periodicity** is thus created by infinite repetition of the same motif.

We may see translational symmetry everywhere: border decorations, parking meters, gutters, water fountains, lamps, columns, trees, soldiers, etc. The simple rule for generating these patterns is to define the basic motif, then repeat it at a certain distance again and again.

Symmetry type
Translational symmetry:
Repeating the same object or motif simply by shifting it a constant distance

Symmetry operation (action)
Translation

Symmetry element (tool)
Constant shift

Periodicity:
The repetitive occurrence of exactly the same motif

OVERLEAF: *Colonnade on St. Peter's Square, Vatican City*

ABOVE TWO PHOTOS: *Walls of the Great Mosque in Cordoba, Spain*

Lights on Alexander III Bridge in Paris, France

Street lamps in Budapest, Hungary

Vincent van Gogh's painting, On Montmartre

Columns in the Palais Royal, Paris, France

Columns surrounding the yard of the Kunjongjon Hall, Seoul, Korea

Kennedy Center, Washington, D.C.

St. Peter's Square in Vatican City

Columns and trees on the campus of the State University of New York at Albany

Israel

Honolulu, Hawaii

Moscow, Russia

Budapest, Hungary

XI. REPEATING EVERYTHING

FAR RIGHT:
Moscow, Russia

RIGHT:
School buses in Storrs, Connecticut

FAR RIGHT:
Aqueduct from Roman times in Sicily, Italy

RIGHT:
Railway terminal in Rome, Italy

Parking meters in Baltimore, Maryland

Gutters in Manezh Square, Moscow, Russia

Water fountain from Roman times, near L'Aquila, Italy

Showgirls in an open-air Japanese television taping, Kyoto, Japan, 1992

Royal Danish guards

Czechoslovakian soldiers (1982)

Benches in Erice, Sicily, Italy

Repetition of a motif can extend to infinity, at least in our imaginations. An important feature of translational symmetry is that, at least in principle, it is not terminated. Thus, whenever we describe translational symmetry here, and later in more complicated cases, we will visualize extension to infinity.

Railroad tracks are ideal examples of patterns going on to infinity. So are fences.

RIGHT:
Liberty Bridge, Budapest, Hungary

FAR RIGHT:
Korea

Fences can demonstrate the economical aspects of translational symmetry. We can use the principle of mass production: we merely design and produce one structural element and then produce as many identical copies as needed.

St. Petersburg, Russia

St. Petersburg, Russia

Korea

Korea

Repetitive Symmetry

So far, we have looked at examples of simple translational symmetry; that is, simple shifting of the basic motif from one position to another position and then to another position, and so on. Translational symmetry is also called **repetitive symmetry**. However, repetitive symmetry has a broader meaning. Repetition may be achieved by other means as well. It is not only simple translation, but also other symmetry operations, such as reflection or rotation, that can be repeated.

The Seven Classes of Band Patterns

As an example, let's choose a black triangle for our basic motif and find all the possibilities for its repetition:

1. Here the triangle is simply shifted a certain distance: This process is simple translation. We have already seen many examples of translation on the previous pages—street lamps, trees, soldiers, and so on.

2. Repetition is achieved by a combination of translation and horizontal reflection.

This combination is achieved as follows:

These steps are then repeated over and over:

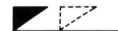

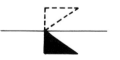

. . . and so on.

The emerging pattern is:

The dashed line (— — — —) on the drawing indicates the presence of this *combined* symmetry element: translation followed by horizontal reflection. This symmetry element is also called a *glide reflection plane.*

Glide reflection plane:
A combined consecutive application of translation and horizontal reflection

3. The repetition at right is achieved by 2-fold rotation of the single triangle motif, as shown below.

● is used to show the presence of the 2-fold rotation axis.

4. Repetition here is achieved by vertical reflection. Horizontal and/or vertical *thick* lines on the drawings indicate the presence of reflection planes (horizontal and/or vertical).

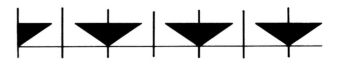

5. Repetition is achieved by horizontal reflection and translation.

6. Repetition is achieved by 2-fold rotation, followed by vertical reflection.

7. Repetition is achieved by alternating vertical and horizontal reflections.

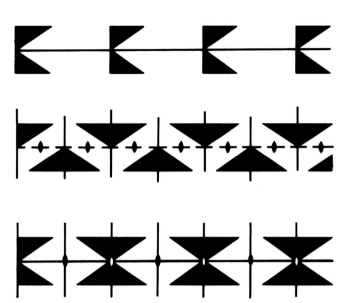

Looking at patterns 6 and 7 we can see that the combined application of symmetry elements to a motif may generate additional symmetries. For example, in pattern 6, the application of vertical reflection and 2-fold rotation generates glide reflection. In pattern 7, the combined application of two kinds of reflection generates 2-fold rotation as well. All symmetry elements (such as reflection planes and rotational axes) are indicated in the drawings.

Here we have shown seven possibilities of how to create one-dimensional repetitive patterns, and curiously, there are no other possibilities.

Let's now consider some practical examples of these patterns.

Below: from one footprint, the next is generated by glide reflection; that is, by a combination of translation and horizontal reflection. Thus, these footprints (and also the oarsmen on the next page) belong to pattern 2 in the series of seven classes shown on these pages.

2.

5. On the other hand, the legs of the centipede can be considered as belonging to pattern 5, a combination of horizontal reflection and translation.

Hungarian Needlework

All seven classes of the band patterns are illustrated here by Hungarian needlework, collected by noted Hungarian folklorist Mrs. Györgyi Lengyel. It may take you a little while to recognize the seven symmetry classes of these and similar patterns, but if you look at the drawings of the small black triangles under each example, the patterns should emerge in your mind.

1. *Edge decoration of tablecloth*

2. *Pillow-end decoration*

3. *Decoration patched on shepherd's felt coat*

4. *Edge decoration of bed sheet*

5. *Decoration of shirt front*

6. *Pillow decoration*

7. *Tablecloth decoration*

On this and following pages are examples of other band patterns, including Greek, Roman, Egyptian, Mexican, Native American, Arabic, Persian, Japanese, and Chinese decorations.

Greek Ornaments

Fret ornament

Fret ornament

Terra-cotta ornament

Fret ornament

Terra-cotta ornament

Roman Border Designs

Border designs in terra-cotta

Egyptian Border Designs

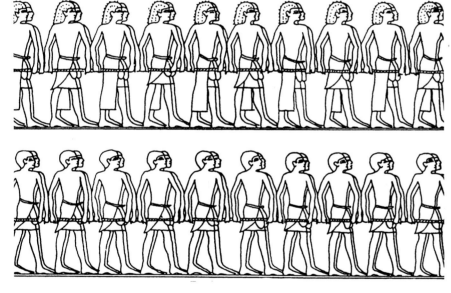

Workers transporting monuments, depicted on tomb

Frieze, Temple of Denderah Tentyris

Mexican Patterns

Cylindrical stamp

Xicalcoliuhqui designs combined with spiral motif, from Mexico City

Native American Designs

Decorations on pottery deserve special mention, because in a way they fulfill the criterion for infinite repetition of patterns; they never end as we rotate the pots.

Xicalcoliuhqui pattern from Oaxaca, Mexico

Pueblo pottery

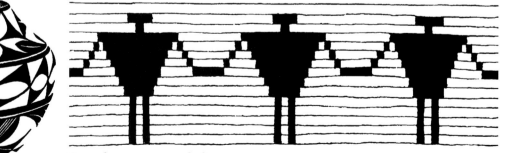

Onondaga wampum belt

Arabic Patterns

Wall tiles in the Mihrab of the Mosque of Cheykhoun (14th century)

Ceramic wall tiles, borders (16th century)

Ceramic wall tiles, borders (16th century)

Ceramic wall tiles from the monastery of the Dervishes (17th century)

Wall mosaic (15th – 16th centuries)

Persian Border Designs

RIGHT & BELOW: *Stucco border patterns from the Masjid-i-Jami in Isfahan, the Masjid-i-Jami in Nayin (10th century) and other buildings*

RIGHT & BELOW: *Stucco border patterns from Varamin, Bostam, and a mausoleum in Qum*

Japanese Border Designs

These designs incorporate recognizable motifs as well as more abstract patterns

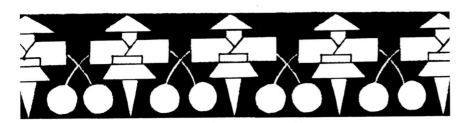

Chinese Lattice Designs

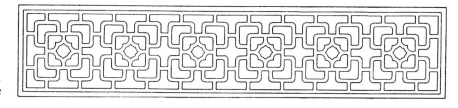

Chengtu, Szechwan Province

*Street balustrade, Han-Line,
Chengtu, Szechwan Province*

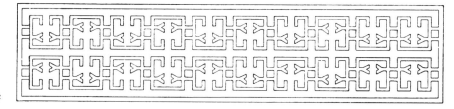

*Balustrade in Buddhist temple in
the mountains, Szechwan Province*

Chengtu, Szechwan Province

*Yü Wang temple, near Shaohing,
Chekiang (Ming Dynasty traditions)*

ABOVE (LEFT AND RIGHT): *Korean beam-end decorations*

Building Decorations

Border patterns are often used to decorate buildings and for mosaic paving patterns.

Ulm, Germany

Prague, Czech Republic

Small Italian island off Sicily, Italy

Italy

143

Papercutting

A simple technique for generating border decorations is papercutting.

Inducing the Feeling of Motion

Border decorations may induce the feeling of motion. They may also convey the feeling of direction. Thus, such decorations may help to move crowds of people in underground passages, railway stations, and similar places without signs telling people explicitly to go in this or that direction.

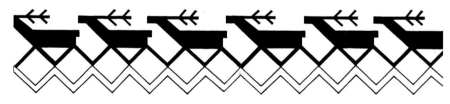

XII. HELIX & SPIRAL

Helices

The border decorations introduced in the previous chapter show the extension of periodic repetition in one direction. With **helices** and **spirals,** there is also repetition in one direction, but the difference is that they are accompanied simultaneously by rotation.

What is the symmetry of a spiral staircase? Is it rotational or translational? It is both. With each step, there is a movement along the axis of the spiral staircase and a small rotation as well. A little translation and a little rotation at the same time, repeated, in principle, to infinity. In reality, of course, all spiral staircases end somewhere, but they need not do so in your imagination, where they can go on to infinity.

Right: *Spiral staircase at a Tel Aviv fire station, Israel*

Far right: *Ottawa, Canada*

Right: *Sicily, Italy*

Far right: *Fukuoka, Japan*

Overleaf: *Spiraling plant* (Euphorbia myrsinites) *in Pécs, Hungary*

Rotation:
When an object is rotated around its axis, it appears in the same position two or more times

Symmetry element (tool)
Axis of rotation

Translation:
Repeating the same object or motif by shifting it a constant distance

Symmetry element (tool)
Constant shift

Seoul, Korea

Kyoto, Japan

To be exact, the spiral staircase—extending to infinity in our imagination—does not have the symmetry of a spiral. Rather, it has the symmetry of a **helix.** This symmetry is characterized by a constant amount of translation accompanied by a constant amount of rotation. The symmetry of a **spiral** differs from that of a helix in that the amount of rotation and translation in a spiral changes gradually and regularly.

Helical symmetry:
Translation accompanied by rotation, with the amount of translation and rotation constant

Spiral symmetry:
Translation accompanied by rotation, but the amount of translation and rotation changes gradually and regularly

Spiral staircase in a bombed-out palace near Potsdam, Germany (photograph taken in 1980)

Frank Lloyd Wright's Guggenheim Museum in New York is itself a huge spiral staircase, although there we have to walk round and round up a ramp, rather than climbing stairs

RIGHT: *The impossible stairway: With a small trick in the drawing, it's possible to give the impression of a stairway on which one can walk around to infinity*

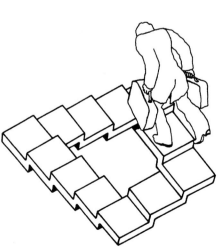

Grape decorations around columns at an Eastern Orthodox monastery in Zagorsk, Russia

The columns in a monastery in Zagorsk, Russia (above), are decorated with grapevines displaying helical symmetry.

The directions of the two helices, however, are different, with mirror symmetry between the two. Thus, helices, as well as spirals, may be left-handed or right-handed—they may be chiral.

On an entirely different scale, many biologically important macromolecules have helical structures, as shown at right.

Spirals

The helix may also be considered a special case of a spiral in which the amount of rotation and translation remains constant. While a helix always extends in three dimensions, a spiral can also be drawn on a piece of paper, that is, in two dimensions.

Chiral:
Describes an object that cannot be superimposed on its mirror image

Helical biological macromolecule

Left-handed and right-handed helices from a textbook on biochemistry

Tropical Storm Sam in the eastern Indian Ocean off the western coast of Australia, as photographed January, 1990, by STS-32 astronauts. The eye of the storm is visible in the center, with swirling bands of the storm rotating clockwise toward the center

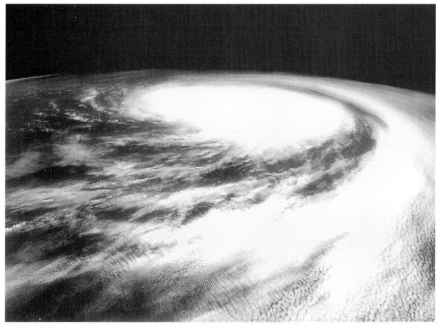

An oblique view of Hurricane Pefa in the Pacific Ocean east of Taiwan, shot in August, 1991, by STS-43 astronauts

In Natural Phenomena

Galaxies, cloud spirals, and water swirls all follow this pattern. An example easily observed in everyday life is the bathtub vortex.

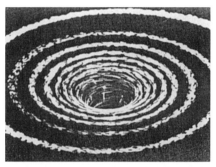

Bathtub vortex formed when water drains out of the bathtub or wash basin

The Whirlpool Galaxy in the constellation Canes Venatici. It is composed of stars, gas, and dust

Clouds in a meteorological report

XII. HELIX & SPIRAL

In Art

Spirals often occur in artistic creations of many varieties.

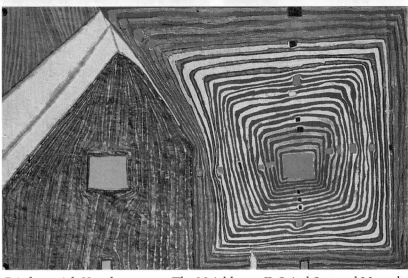

Friedensreich Hundertwasser: The Neighbours II: Spiral Sun and Moon-house

The Spiral Is the Symbol of Life and Death

This spiral lies at that very point where inanimate matter is transformed into life.

I am convinced that the act of creation took place in form of a spiral.

Our whole life proceeds in spirals. Our earth describes a spiral course. We move in circles, but we never come back to the same point. The circle is not closed. We only pass the same neighbourhood many times. It is characteristic of a spiral that it seems to be a circle but is not closed.

The true spiral is not geometric but vegetative. She has swellings, becomes thinner and thicker and flows around obstacles who are in her way.

The spiral shows life and death in both directions. Starting from the center, the infinite small, the spiral means birth and growth, but by getting bigger and bigger the spiral dilutes into the infinite space and dies off like waves who disappear in the calm waters.

William Blake, Jacob's Dream

On the contrary if the spiral condenses from outer space, life starts from the infinite big, the spiral becomes more and more powerful and concentrates into the infinitely small which cannot be measured by man because it is beyond our conception and we call it death.

The spiral grows and dies like a plant—the lines of the spiral, like a meandering river, follow the laws of growth of a plant. It takes its own course and goes along with it. In this way the spiral makes no mistakes.

F. Hundertwasser, 1991

Vincent van Gogh, Starry Night in
St. Rémy

Computers can be used to create spiral wonders with relatively simple programs.

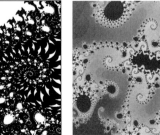

*Computer drawings of spirals
by Clifford A. Pickover*

Shells

LEFT: *Fossil snails exhibited in a park in Vaduz, Lichtenstein. They were split down the middle, so that—just as positive and negative—the two photos have an antisymmetric relationship*

Right-handed and left-handed shells

Shells along the Texas coast

A species of solarium

Native California snail

Life Forms

Spirals occurring frequently in animals and plants indicate an underlying principle of mathematical control in certain aspects of life forms.

Tibetan shawl goat

RIGHT:
Bighorn sheep

Greater kudu, central Africa

Tendrils of plants usually grow in long spirals, as shown here by the tendrils of the wild cucumber. The stalk of the storksbill fruit is straight when wet, but begins to twist into spirals as it dries.

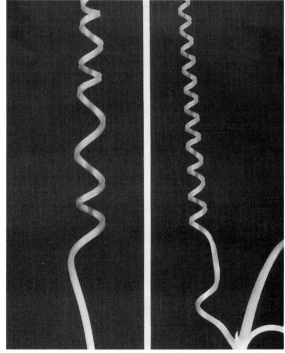

Stalk of storksbill fruit   *Tendrils of wild cucumber*

Tower of Babel, as painted by Pieter Bruegel the Elder, in 1563—the tower was a spiral with seven terraces in the ancient city of Babylon

Towers

Towers frequently exhibit spiral symmetry.

The Galilei Tower of the Heureka Exhibition, Zürich, Switzerland, (1991)

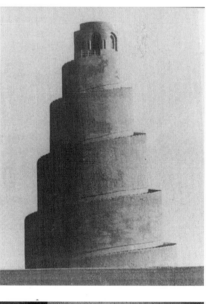

Malwiya, the Great Mosque, Samarra, Iraq, 9th to 15th centuries

Towers in Copenhagen, Denmark

Tatlin's design for a monument of the Third International

The Fibonacci Numbers

Perhaps the most beautiful occurrence of spiral symmetry in nature is the scattered leaf arrangement around the stems of plants, a phenomenon botanists call *phyllotaxis.*

The stem of *Plantago media* certainly does not extend to infinity. However, if we take some philosophical liberties, we may consider the plant/seed/plant/seed/plant . . . sequence to extend to infinity, over a period of time. Thus, the leaf arrangement of a single stem could be thought of as part of an infinite series.

Let's now consider the relative positions of the leaves around the stem of *Plantago media.* Starting from leaf 0, circle the stem looking for the next leaf that would be exactly above the initial leaf. This will be leaf 8, and we discover that we have to circle the stem three times before we reach it. The ratio of the two numbers is 3/8, and this tells us that a new leaf occurs at each 3/8 part of the circumference of the stem.

The simplest case is when the leaves occur on opposite sides of the stem as we move along it. An example is the leaf arrangement of the simple yellow flower *Oenothera biennis.*

Again we start with a leaf labeled 0, and circle the stem until we find another leaf exactly eclipsing the leaf 0. This will be leaf 2, after one complete circle around the stem. The ratio of the two numbers in this case is 1/2, telling us that a new leaf is always found at half of the circumference of the stem.

ABOVE: Plantago media

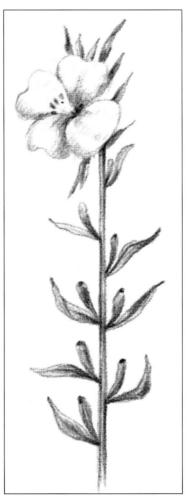

Oenothera biennis

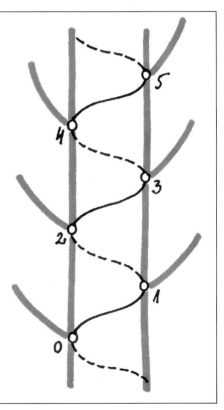

It has been suggested that having successive leaves on a stem separated in this way maximizes solar illumination and air for the leaves.

The two cases demonstrated are characterized by 3/8 and 1/2, respectively. Other leaf arrangements for various plants:

Plant	Circlings/leaf numbers
Common grasses, elm tree, basswood tree	1/2
Sedges, beech tree, hazel tree	1/3
Most fruit trees, oak tree	2/5
Plantains, poplar tree, pear tree	3/8
Leeks, willow tree, almond tree	5/13

The numbers of necessary circlings and the leaf numbers can be arranged in separate series:

Circlings: 1, 1, 2, 3, 5, etc.

Leaves: 2, 3, 5, 8, 13, etc.

In both these series, *each number is the sum of the previous two numbers:*

$$1+1=2$$
$$1+2=3$$
$$2+3=5$$
$$3+5=8$$
$$5+8=13$$
$$8+13=21$$
$$13+21=34$$
$$21+34=55, \text{etc.}$$

The two series, that is, the series of circlings and the series of leaves, can be joined, and this can extend to infinity:

1 1 2 3 5 8 13 21 34 55 89 144, . . . etc.

This number series is called the **Fibonacci series** after its discoverer, Leonardo of Pisa (Fibonacci), an Italian mathematician who lived in the 13th century.

Hanging heliconia (Heliconia collinsiana), *Hawaii*

"Human subtlety . . . will never devise an invention more beautiful, more simple, or more direct than does nature, because in her inventions nothing is lacking, and nothing is superfluous."

Leonardo da Vinci
The Notebooks (1508 – 1518)

XII. HELIX & SPIRAL

In Leaves and Plants

Beautiful examples of spiral leaf arrangements abound in plant life around the world.

Euphorbia myrsinites, *Pécs, Hungary*

Cactus

RIGHT: Echeveria, *Pécs, Hungary*

Ginger, Hawaii

Prickly cycad, Hawaii

Echium, *California*

Pineapple, *Hawaii*

156 SYMMETRY

Palm tree, Israel

Palm trees, Texas

Hawaii

Stalk of elephant ear
(Kalanchoe beharensis), *Hawaii*

Brussels sprouts, Pécs, Hungary

In the botanical garden,
Madrid, Spain

XII. Helix & Spiral

Pine cones

The Fibonacci numbers also occur in the numbers of the spirals of scales of pine cones observed from below. There are 13 left-bound spirals of scales and 8 right-bound spirals, both Fibonacci numbers.

Much larger Fibonacci numbers can be observed in the left-bound and right-bound spirals of the seed arrangement of daisies and sunflowers, the spikes of a cobweb thistle, as well as the florets of a cauliflower. Both the pine cone scales and the sunflower seeds can be considered as if they were compressed leaf arrangements around their stems. Thus, the relationship to the previous plants is obvious.

Sunflowers

Singapore stamp with a daisy

Cauliflower

Right: *Cobweb thistle* (Cirsium occidentale), *California*

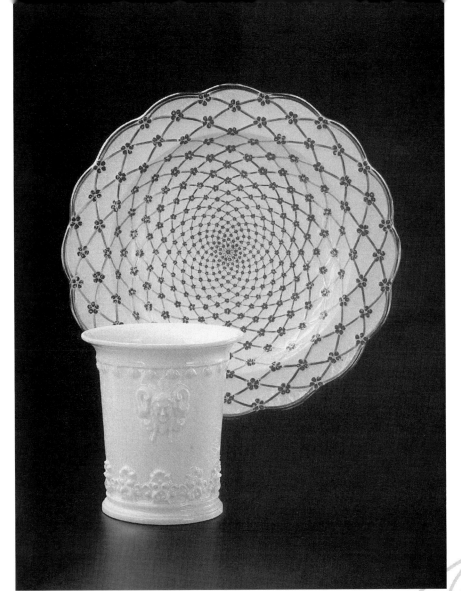

In Decorative Design

Humans have adapted analogs of spiral forms throughout history in a wide variety of artistic creation.

LEFT: *A Russian porcelain plate made around 1760 in St. Petersburg for the personal use of Empress Elizabeth I, daughter of Peter the Great*

BELOW: *White Mountain Apache design (Native American)*

Michelangelo's design of a continuous pathway in the quadrangle of the Capitol, Rome, Italy. (From an engraving by Du Pérac in 1569)

The Golden Ratio

Let's now look at the ratios characteristic for phyllotaxis (the arrangement of leaves around a stem), but this time consider the actual values for these fractions (with each step, the ratio more closely approximates the golden proportion):

$$1/2 = 0.500$$

$$1/3 = 0.333$$

$$2/5 = 0.400$$

$$3/8 = 0.375$$

$$5/13 = 0.385$$

$$8/21 = 0.381$$

$$13/34 = 0.382$$

$$21/55 = 0.382$$

$$\vdots$$

$$= 0.381966\ldots$$

This extremely important, so-called irrational number expresses the golden ratio, which, in turn, is derived from the golden section.

The Golden Section

The **golden section** (also called the Divine Proportion) was said by Kepler to be "one of the two treasures of geometry," and was considered by Plato (in *Timaeus*) as the key to the physics of the cosmos. This mathematical relationship appears repeatedly in growth patterns in nature and has fascinated mathematicians and artists for centuries.

What is the golden section? It means that a certain length is divided in such a way that the ratio of the longer part to the whole is the same as the ratio of the shorter part to the longer part. In this case, if the whole is unity—that is, 1.000—then here you will divide 1.000 into two parts, one of the length 0.618, and the other 0.382.

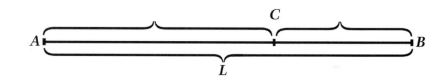

Line AB is divided at C so that:

The ratio of AC to AB is the same as the ratio of CB to AC.

Ceiling decoration in the Hermitage, St. Petersburg, Russia. (Note that the repeated motif is a double-headed eagle)

The Golden Rectangle

There is a special rectangle with proportions corresponding to the golden ratio. It is called the **golden rectangle.**

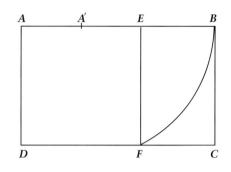

It is not difficult to construct such a rectangle. You will need a pencil, ruler, compass, and a right-angle triangle. First draw a square, *AEFD*, of arbitrary size. Then divide the line *AE* in half at *A'*.

Then, with the compass and using *A'* as center, draw an arc from *F* up to *B*, which intersects the extension of line *AE* at *B*. With your triangle, draw *BC* perpendicular to *AB*, meeting the extension of line *DF* at *C*. The new *ABCD* rectangle is a golden rectangle, in which *AB* is divided by *E* in exactly the golden section:

$$AE{:}AB = EB{:}AE$$

That is, the ratio of the longer part to the whole is equal to the ratio of the shorter part to the longer part. This is why *E* is called the "golden cut."

The Logarithmic Spiral

The construction of the golden rectangle is interesting also because it can show the connection among the golden section, spirals, and the Fibonacci numbers. To demonstrate:

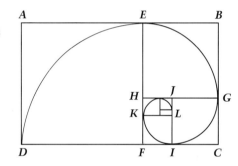

- Take the same rectangle *ABCD* and draw through its golden cut *E*, the line *EF*, which is perpendicular to *AB* and cuts off the square *AEFD* from the rectangle. The remaining rectangle *EBCF* is also a golden rectangle.

- Continue cutting off the squares from within these golden rectangles.

- Cut off the square *EBGH* from rectangle *EBCF*. This leaves the new smaller golden rectangle *GCFH*.

- Then, from this cut off the square *GCIJ*, leaving the smaller golden rectangle, *IFHJ*.

- Next, from this rectangle, cut off the square *IFKL*, which leaves the golden rectangle *HJLK*, and so on. You can continue this process—at least in your imagination—indefinitely, until a rectangle, indistinguishable from a point, is reached.

Then take a compass and from the inside corner of each square (*F*, for example, for square *AEFD*), draw an arc from one corner of the square to the diagonally opposite corner (*D–E*, for example). Then from *H*, draw arc *E–G* and so on. This procedure will give you a spiral.

Logarithmic Spiral and Golden Section

There are several interesting mathematical relationships between the features of the golden rectangle and the spiral that show the connection between the spiral and the golden section. This spiral has been called by different names corresponding to one or another of its characteristics. Descartes called it the *equiangular spiral,* Halley called it the *proportional spiral,* and Bernoulli used the phrase *logarithmic spiral.*

There is also a connection between the spiral and the Fibonacci series. The spiral passes through diagonally opposite corners of successive squares (*D–E, E–G, G–I, I–K,* etc.). The lengths of the sides of these squares form a Fibonacci series. If the smallest square (not shown here) has a side of length *a,* the adjacent square also has a side of length *a.* The next, third, square has a side of length *2a,* the next *3a,* followed by lengths *5a, 8a, 13a* and so on, which is the Fibonacci series.

A beautiful feature of the logarithmic spiral is that although two segments of the curve have different *sizes,* their *shape* is always the same. If we take a smaller segment of this spiral and enlarge it on a copy machine, it can be brought into exact coincidence (fit) with a larger-size segment of the curve. The spiral does not have a terminal point. It can grow indefinitely, but its shape remains unchanged.

This fundamental property of the logarithmic spiral corresponds precisely to the biological principle that governs the growth of many shells. The principle is the simplest possible: the size increases, but the shape remains the same. The only mathematical curve to follow this pattern of growth is the logarithmic spiral. This is why Jacob Bernoulli described it as *spiral mirabilis* (miraculous spiral) [*Acta Eruditorum,* 1691].

With the mathematical knowledge just gained, we can now return to our examples from nature with a more informed perspective. Note that the successive chambers of the nautilus seashell follow the form of the logarithmic spiral exactly. As the shell grows, the size of the chambers increases, but their shape remains the same.

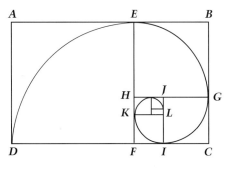

In the growth of a shell, we can conceive no simpler law than this, namely, that it shall widen and lengthen in the same unvarying proportions: and this simplest of laws is that which Nature tends to follow. The shell, like the creature within it, grows in size, but does not change its shape; and the existence of this constant relativity of growth, or constant similarity of form, is of the essence, and may be made the basis of a definition, of the equiangular spiral.

D'Arcy W. Thompson
On Growth and Form

Leonardo da Vinci also recognized the principle at work, and wrote:

The creature that resides within the shells constructs its dwelling with joints and seams, and roofing, and other various parts, just as a man does in the house which he inhabits; and this creature expands the house and roof gradually in proportion as its body increases and as it is attached to the sides of these shells.

From T. A. Cook
The Curves of Life

Both the Fibonacci numbers and the golden ratio seem omnipresent in nature. The two are shown to be intimately related, since the golden ratio is obtained when we take fractions of very large Fibonacci numbers. You might well ask, What is the relation to symmetry, since the golden ratio seems to be so asymmetrical? There are two connections. One is that when patterns can be generated by simple rules, there is a kind of symmetry. The other (as we discussed in the introduction) is that a broader definition of symmetry would include harmony and proportion, and the golden ratio is certainly abundant in these qualities.

Proportions

Shown here are two famous
examples of the golden ratio,
Michelangelo's painting of
Adam's Creation in the Sistine
Chapel, and the Bauhaus
building. The proportions of
Michelangelo's painting are
indicated by the architect
G. Doczi. They represent the
golden ratio. Bauhaus designs
were famous for their
proportions.

G. Doczi, The proportions of Adam's
Creation *by Michelangelo (Sistine
Chapel, Vatican City), all represent-
ing the golden ratio, are indicated*

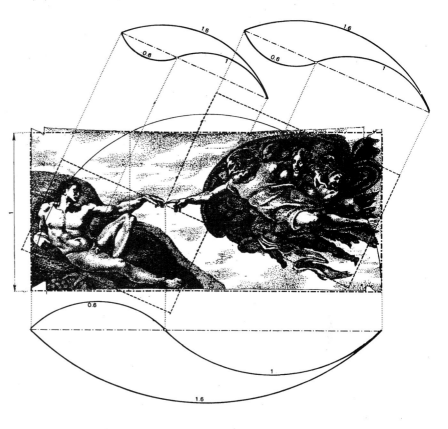

*East
German
stamp*

*The Bauhaus building itself, built
in 1926—Dessau, Germany*

Patterns from Circles

Earlier, in the chapter on repetition, we saw how to generate a pattern from a motif by translation or by other symmetry operations such as reflection or rotation, followed by repetition. We saw a great variety of endless patterns repeating in one direction. Now we are going to see the extension of repetition in two directions and the creation of planar repetitive patterns. In other words, we are going to talk about covering a plane.

To build a repetitive two-dimensional planar pattern, first select a motif. For example, take a shape—the circle.

First, repeat this motif in one direction by simple translation. This produces an endless row of circles.

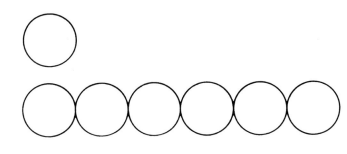

Next, repeat this row many times over. This way we get an endless network in two directions, that is, in a plane. Here, it is illustrated by Korean designs. The pattern on the right shows more efficient packing.

Repetition:
Repeated application of the same symmetry operation—reflection, rotation, or simple translation

Symmetry element (tool)
The tool corresponding to whatever operation is repeated (**mirror plane** for reflection, **axis of rotation** for rotation, **constant shift** for translation)

Translation:
Repeating the same object or motif by shifting it a constant distance

Symmetry element (tool)
Constant shift

Periodicity:
The repetitive occurrence of exactly the same motif

OVERLEAF: *Worker bees and their honeycomb*

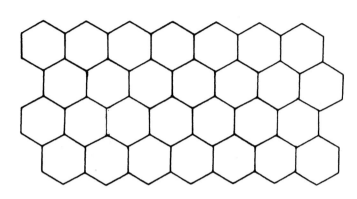

Patterns from Hexagons

This network of equal-sized hexagons is periodic (repeating the same motif). It is periodic because all the hexagons are equal in size. Moreover, it is periodic in two directions because the hexagons cover the whole plane.

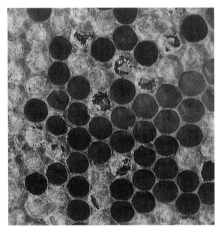

To build a honeycomb from wax, bees first form a network of closely packed circles. The bees are near equal size and move around in circles, creating circles in the wax. Although the circles are as closely packed as possible, they do not cover the available surface completely. The liquid wax flows into the spaces between the circles and forms hexagons. The hexagons then cover the entire available surface without gaps.

Newly born bees starting to emerge from cells

Worker bees create wax for combs from secretions of tiny abdominal glands

Worker bees

Wax comb of honeybees: a masterpiece of art and engineering

Base of offshore oil platform after it is turned upside down

Oil platform being towed to destination before being turned upside down

Hexagonal Designs, Human-made and Natural

This picture shows what looks like a honeycomb. Actually, it is a concrete base under construction for an offshore oil platform in the North Sea. The base consists of a network of regular hexagonal shapes, similar to the honeycomb.

The symmetry of the oil platform as well as the symmetry of the honeycomb come from repetition in two directions. What is the most important difference between the network of circles and the network of regular hexagons? The regular hexagons cover the whole surface without gaps or overlaps, while with the circles, a lot of space is left in the gaps between the circles.

The moth's compound eye, (magnification x 2000), also shows hexagonal subdivision

Columnar basalt joints show a hexagonal division of the surface

Fish scales on an Australian stamp

All these are examples of approximate arrangements. (The hexagons are not perfect.)

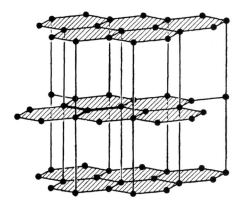

Model of graphite structure—the filled circles indicate the carbon atoms

The structure of graphite layers, on the other hand, shows a virtually perfect system of closely packed regular hexagons.

Covering the Surface with Regular Polygons

Curiously, the only *regular* polygons (equal sizes) that can cover a surface without gaps or overlaps are the equilateral triangle, the square, and the regular hexagon.

If we try to cover a flat surface with, for example, regular pentagons of equal size, there will always be some gaps, no matter how we arrange the pentagons. There are always some rhombi (equilateral parallelograms) left between the pentagons.

The same is true for regular octagons (eight-sided figures) of equal size in that they cannot completely cover the available surface. Here, there are always small square areas left uncovered between the octagons.

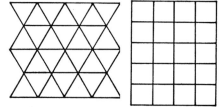

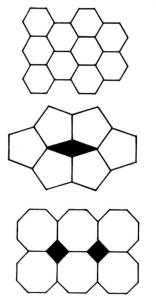

Regular polygon:
All its angles are equal and all its sides are of equal length

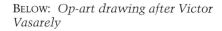

BELOW: *Op-art drawing after Victor Vasarely*

LEFT: *Hungarian needlework pattern*

169

Similarly, none of the higher regular polygons, such as the regular nonagon (nine-sided figure), regular decagon (ten-sided figure), etc., can cover a flat surface completely.

Although only three of the regular (and same-size) polygons can cover a surface without gaps and overlaps (equilateral triangle, square, equilateral hexagon), there is an unlimited number of irregular polygons and other arbitrary shapes that can do this.

We will explore some techniques for their construction in the next chapter, Rhythm on the Wall.

Patterns from Pentagons

If the stipulation for equal size is relaxed for the regular polygons, and thus the pattern is no longer periodic, it is possible to create patterns covering the whole surface without gaps or overlaps. An example is the pattern created from regular pentagons of gradually changing size.

Take seven regular pentagons and combine six of these to make a large pentagon, as at far right:

Take the seventh pentagon and divide it thusly:

Take the five triangles generated from the seventh pentagon and use them to fill in the gaps in the large pentagon. Then take six more large pentagons and continue the procedure we just described.

This construction is then repeated on an ever-increasing scale. The hierarchic packing of pentagons builds up into an endless regular though nonperiodic network illustrated by a computer drawing.

Note the small regular pentagon left over in the center

Note the triangular gaps in between the edges

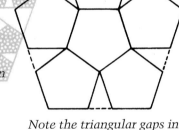

Pentagonal snowflake

This pattern is the result of the construction

Planar Patterns

The honeycomb, the graphite structure, the oil platform, and the moth's compound eye are examples of planar patterns. From any one-dimensional pattern with periodicity, it is easy to generate a planar pattern by repetition, extending the periodicity in two directions.

Here are two movie billboards in Madrid, Spain. Now extend these patterns to infinity in your imagination.

There are countless planar patterns around us; that is, patterns extending in two dimensions. The basic motif is repeated not just in a row but in the whole plane. It may be the decoration of a summer dress, the pattern of a veil, a field planted with trees, the arrangement of stones on the pavement, tiling, a brick wall, a brick chimney, roofing tiles, the roof structure of a basketball court, a parquet floor, fences, wallpaper designs, Andy Warhol paintings of Campbell's soup cans or Coca-Cola bottles, and so on, and on and on. Each of these patterns can be looked at as part of an endless network generated by some simple rules, and if we do so, we can describe their symmetry.

On either side of the billboard is a border decoration

The same border decoration is extended into a planar pattern next to this billboard

Repetition:
Repeated application of the same symmetry operation—reflection, rotation, or simple translation

Symmetry element (tool)
The tool corresponding to whatever operation is repeated (**mirror plane** for reflection, **axis of rotation** for rotation, **constant shift** for translation)

Periodicity:
The repetitive occurrence of exactly the same motif

OVERLEAF: *Sophia Loren behind a veil. Photo: Denis Taranto,* Jet-Set

Javanese batik designs

Textile designs

Portuguese tiling, Lisbon, Portugal

Roof structure of a gymnasium in Storrs, Connecticut, under construction, 1988

Main square, Baja, Hungary

Street pavement in Erice, Sicily, Italy

Paving pattern in Annapolis, Maryland

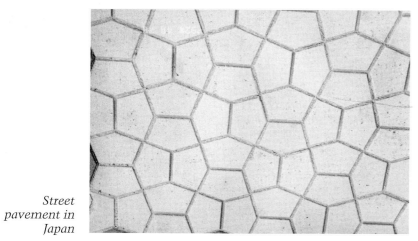

Street pavement in Japan

L'Aquila, Italy

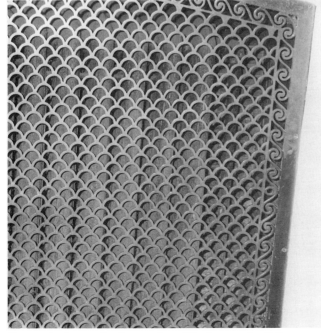

Andy Warhol, Campbell's Soup Cans

Seats at the Olympic Stadium in Seoul, Korea

Fence in Taejon, Korea

Screen in front of a heater in a Manhattan building

Andy Warhol, Coca-Cola Bottles

XIV. RHYTHM ON THE WALL

Church roof in Budapest, Hungary

Roofing of village house, France

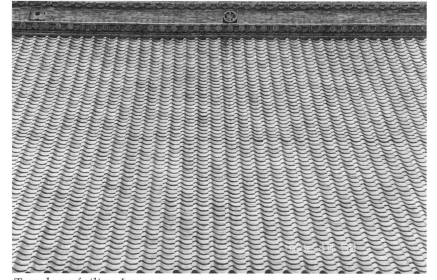

Temple roof tiling, Japan

Brick wall, Moscow, Russia

Pavement, Jaen, Spain

Tile roof, Granada, Spain

Fence and pineapple field in Oahu, Hawaii

Tea plantation in Fukuroi, Shizuoka, Japan

Creating Planar Patterns

One easy way to make a pattern is on graph paper with squares. We choose a motif that is first repeated in one direction to create a row. This may be done by simple translation or by applying symmetry elements like reflection planes or rotation axes. (This is the way we made the patterns described on pages 131–133.) The row is repeated to make a planar network. Some very nice patterns emerge, especially if we ignore the underlying network of squares.

Let's select a simple motif, such as a quarter of a circle:

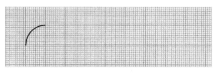

Then repeat it in a row:

Then repeat the row to make the planar pattern:

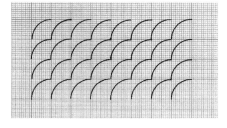

Finally, erase the network:

Now, let's start with the same motif, but apply 2-fold rotation to create the one-dimensional pattern:

Now repeat the same row:

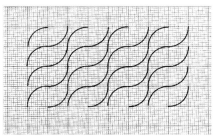

Finally, erase the network:

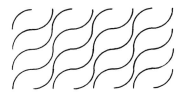

Rotation:
When an object is rotated around its axis, it appears in the same position two or more times

Symmetry element (tool)
Axis of rotation

Translation:
Repeating the same object or motif by shifting it a constant distance

Symmetry element (tool)
Constant shift

Now, take the same row, but instead of simply repeating it, apply a horizontal mirror to produce the next row, and so on:

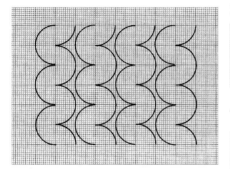

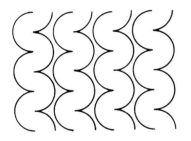

Another possibility is to start again with a quarter circle and apply a glide reflection plane to produce the one-dimensional pattern:

Then repeat this row:

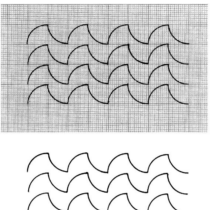

Or use a horizontal reflection plane to produce the planar network:

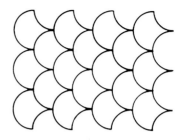

Or rotate the row 180° (2-fold rotation) to make the planar pattern:

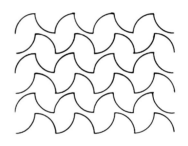

There are countless possibilities.

Reflection:
Reflecting one-half of an object reconstructs the image of the whole object

Symmetry element (tool)
Mirror plane:
Applying a mirror plane to either of two halves, the whole is recreated

Glide reflection:
A combined consecutive application of translation and horizontal reflection

Filling the Surface Completely

Some of the most attractive patterns are those that fill the whole surface without gaps between the motifs or without the motifs overlapping.

Start with a network of identical parallelograms (in pencil):

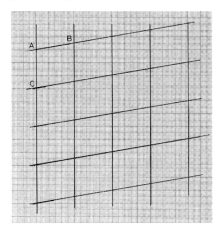

Then connect point *A* to point *B* with an arbitrary line (in ink). Keep this line within the confines of one parallelogram:

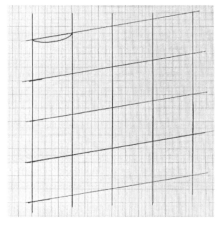

Then connect point *A* to point *C* the same way with another arbitrary line:

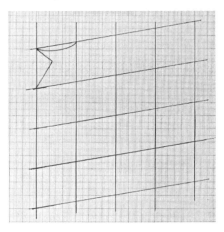

Repeat these two lines in all the parallelograms:

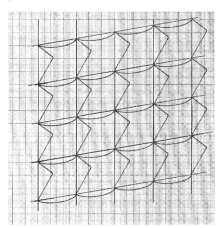

Erase the underlying network of pencil lines:

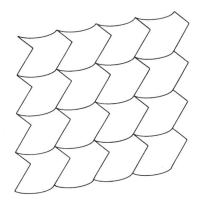

You can also do this starting with any other network of identical polygons.

A Canadian scientist, François Brisse, was so fascinated by such networks that he designed one for each of Canada's provinces and territories. He used the symbol of the provinces as the basis for each drawing. To this basic motif he then applied various symmetries, just as we described for the quarter circle designs.

The Northwest Territories of Canada, for example, has a polar bear as its symbol. A stylized polar bear was chosen as the basic motif. It was first rotated, and the resulting double bear was then repeated in two directions.

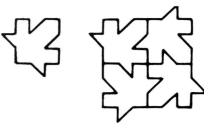

Brisse made a drawing for the whole of Canada as well, starting from a maple leaf. The maple leaf is Canada's national symbol, and he used a simplified shape to create his network. The maple leaf was first rotated to produce a unit of four leaves of 4-fold rotational symmetry. This unit was then repeated in two directions to cover the whole surface.

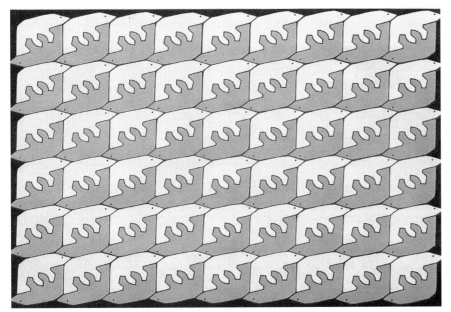

BELOW: *Similar patterns appear to be quite common, in this Portuguese tile, for example*

Decorative Patterns

Such patterns have been used by artists and artisans for hundreds of years. Now that you know something about them, you can look for interesting repeating decorations when visiting ancient places.

In the old town of Badra in the Caucasian Mountains there is a building shaped like a cylinder. It is decorated by a mosaic displaying the word *Allah* some 200 times. This word covers the whole surface of the building.

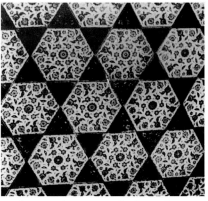

Shibam-Kawkaban, Yemen

Here is the basic motif:

It is then rotated like the maple leaf, and a unit with 4-fold symmetry emerges:

Topkapi Palace, Istanbul, Turkey

This unit is then repeated in two directions to form a planar pattern:

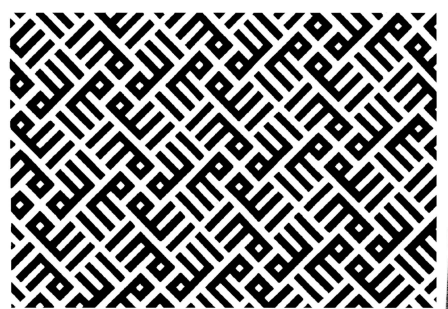

Islamic decoration in Badra, Azerbaijan

Decorations from Arab mosques and the famous Alhambra building in Granada, Spain (facing page) are conspicuously beautiful examples.

Sidi bu Medien, Tlemcen, Algeria

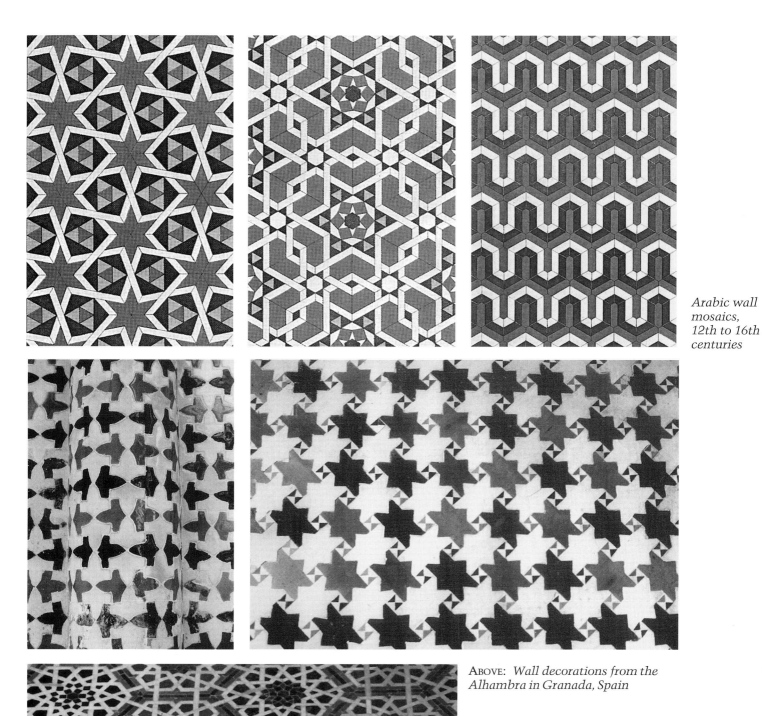

Arabic wall mosaics, 12th to 16th centuries

ABOVE: *Wall decorations from the Alhambra in Granada, Spain*

Floor tiles, church in Palermo, Italy

Seventeen Symmetry Classes for Planar Patterns

Just as there were 7 possibilities for one-dimensional (repeating in one direction only) border decorations *(see pp. 131–133)*, there are exactly 17 symmetry variations for two-dimensional (repeating in two directions) planar patterns. Here, all 17 are shown by examples of Hungarian needlework, preceded by the corresponding patterns derived from the black triangle motif:

TOP AND CENTER: *Patterns of indigo-dyed decorations on textiles for clothing. Sellye, Baranya County, Hungary, 1899*

Indigo-dyed decoration with palmetto motif

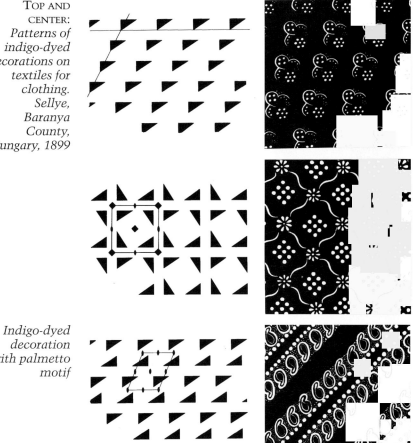

Bird motifs from peasant vests. Northern Hungary

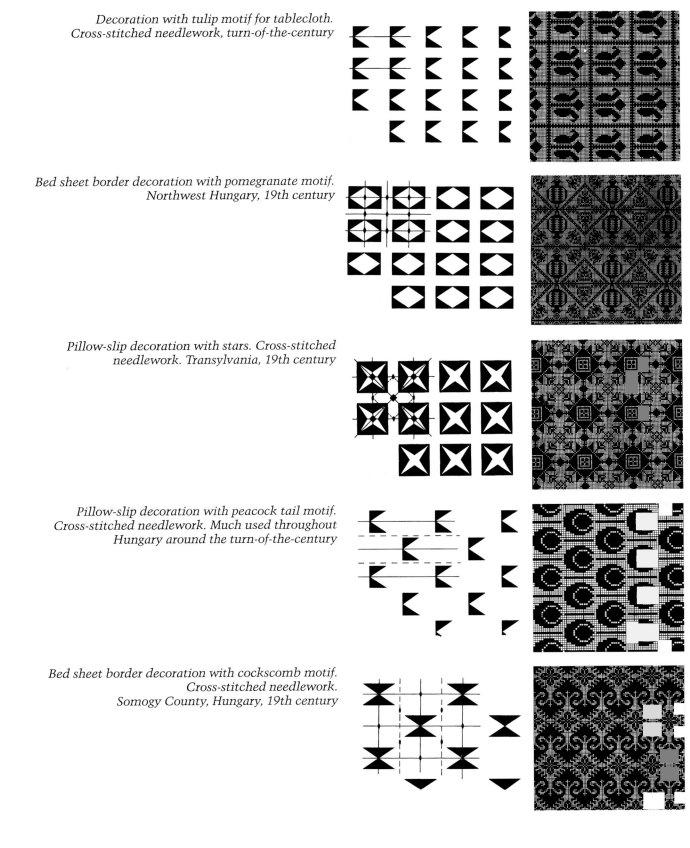

Decoration with tulip motif for tablecloth.
Cross-stitched needlework, turn-of-the-century

Bed sheet border decoration with pomegranate motif.
Northwest Hungary, 19th century

Pillow-slip decoration with stars. Cross-stitched
needlework. Transylvania, 19th century

Pillow-slip decoration with peacock tail motif.
Cross-stitched needlework. Much used throughout
Hungary around the turn-of-the-century

Bed sheet border decoration with cockscomb motif.
Cross-stitched needlework.
Somogy County, Hungary, 19th century

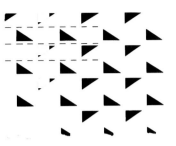
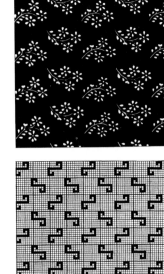

Indigo-dyed decoration. Pápa, Veszprém County, Hungary, 1856

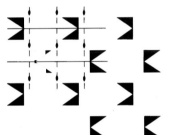

Children's bag decoration. Transylvania, turn-of-the-century

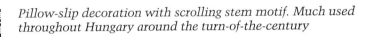

Pillow-slip decoration with scrolling stem motif. Much used throughout Hungary around the turn-of-the-century

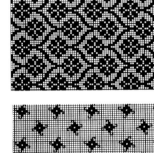

Blouse-arm embroidery. Bács-Kiskun County, Hungary, 19th century

Native American Designs

Thompson butterfly design
basis, British Columbia

Thompson "necklace of bead"
design basket, British Columbia

Thompson "fish net" design
basket, British Columbia

Winnebago thunderbird design
bag, Eastern Woodlands

Winnebago geometric design
bag, Eastern Woodlands

Winnebago bag design with deer and
thunderbird, Eastern Woodlands

Decorations

Tilework designs by Mirza Akbar, early 19th century

Woven carpet, Bagdad, 1396

Patterns in the Onyang Folk Museum

These two Korean patterns are related to each other by antisymmetry (see p. 116)

Artistic Patterns

In this pattern, titled *"Girls,"* the basic motif being repeated is, of course, a girl. There are alternating rows of these girls. In every other row they are standing on their heads with the colors reversed, except for their faces. Here again, we have some antisymmetry in addition to repetition.

This interesting drawing was prepared by an Azerbaijani scientist, Khudu Mamedov. Both Brisse (see p. 181) and Mamedov are crystallographers. Perhaps this is no accident. Crystallographers study the outer shape and the internal structure of crystals. Crystals are built from atoms and molecules by endless repetition in three dimensions. This is what makes them so symmetrical (as you will see in the next chapter, Diamonds & Glass).

Another of Mamedov's drawings is called *Unity*. It seems as if there is a message from history here. The old men are chained and on their knees, and the young are proudly standing. In spite of the overall uniformity, there are different expressions on their faces, especially on the old men's faces.

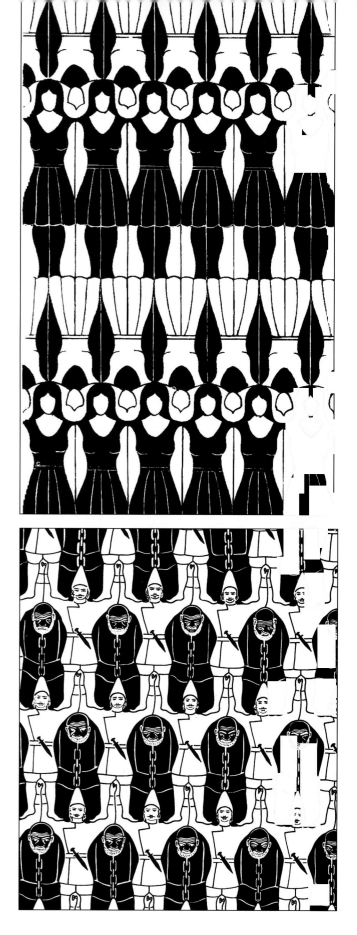

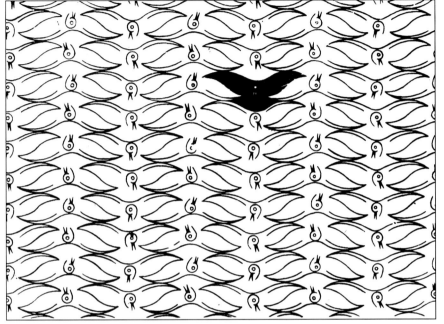

Here is another Mamedov drawing. This one shows sea gulls. First a sea gull is rotated 180° and then the pair is repeated in each row. The rows are repeated in such a way that each bird's head is tossed in alternate directions below the other's.

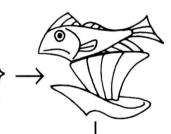

The most famous drawings in this mode were created by the Dutch graphic artist Maurits C. Escher. Here, the basic motif is a fish and a boat. This pair is repeated in two directions to cover the whole surface. (M. C. Escher, *Symmetry Drawing E 113—1962.*)

The motif of this Escher drawing is formed from a bird and a fish. The pair is first rotated by half a turn, and this unit of the two pairs is then repeated in two directions. (M. C. Escher, *Symmetry Drawing E 115—1963.*)

Another of Escher's drawings contains four kinds of animals: a falcon, a fly, a butterfly, and a bat.

The basic unit is a square that consists of one-half of all four animals. This square unit is then repeated by mirror reflections to cover the whole plane. (M. C. Escher, *Symmetry Drawing E 81—1950.*)

Various degrees of abstraction enliven artistic expressions. An interesting example of repetition with considerable variations is the picture named *Homage to Greco* by Hungarian painter István Orosz. El Greco's famous *Study head* is repeated in two directions, each with a different artistic style.

Quite often, gradually changing planar patterns express some process in a powerful way. On the Israeli stamp below, the transformation occurs from the pattern of flowers to the pattern of regular hexagons, while 6-fold symmetry is retained throughout.

LEFT: *István Orosz,*
Homage to Greco

BELOW:
El Greco,
Study head

Ad in the International Herald Tribune, *June 26, 1991*

In an ad calling for protection of the African elephant, the pattern of elephants (on the left) displays translational symmetry in two directions. As we move from left to right, the elephants are gradually turned onto their backs, and eventually only their ivory tusks remain. The symmetry of the pattern is preserved throughout.

Dancing with Symmetry

First with single motifs, and then with border decorations, we have seen that a sense of motion can be conveyed if certain symmetries are present or absent. This is even more true with decorations covering a whole surface.

Joy

The ones shown here, for example, have rotational symmetry only. There is no symmetry plane in them, only rotation. They give us a feeling of rotation, of circling. Patterns such as these may even make us want to dance around! These patterns might make good decorations for the walls of a dancing hall.

Confusion

Consecutive translation and horizontal reflection together is called glide reflection. It is thought to induce the feeling of confusion, so we must be careful when and where we use such decorations. These patterns contain translation and reflection in endless repetition.

Calmness

If we want to have an important meeting, if calmness and respectability are needed, we had better choose a hall decorated with patterns having plenty of symmetry planes, and, preferably, no rotational symmetry.

Capitol Dome, Washington, D.C.

Facades

Modern skyscrapers tend to have uniform facades. They easily lend themselves to very simple, sometimes boring patterns.

LEFT: *Chicago, Illinois*

Houston, Texas

Houston, Texas

Yale Library, New Haven, Connecticut

XIV. Rhythm on the Wall

ABOVE: *Dallas, Texas*

RIGHT: *Building at the Technion
(Israel Institute of Technology), Haifa*

Facades needn't be boring. A truly
unusual pattern was created by the
Viennese artist F. Hundertwasser
for an apartment building in
Vienna. There is no symmetry in it
whatsoever. However, with our
magnanimous approach, we can see
a two-dimensional pattern of more
or less similar individual units.

*Apartment complex designed by
F. Hundertwasser, Vienna, Austria*

XV. DIAMONDS & GLASS

From William Scoresby's log book, 1806

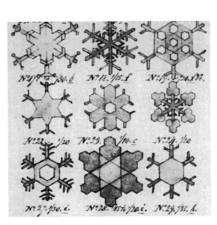

Crystals

The snowflake *(see p. 69)* is a water crystal. The word crystal comes from the Greek *krystallos*, meaning "clear ice." The name originated from the mistaken belief that the beautiful transparent quartz stones found in the Alps were formed from water at extremely low temperatures. By the 17th century, the name crystal was applied to other solids as well. Crystals generally have beautiful symmetrical shapes.

Minerals

Crystals have always fascinated people. Karel Čapek, the Czech writer, wrote the following after his visit to the mineral collection at the British Museum:

There are crystals as huge as the colonnade of a cathedral, soft as mould, prickly as thorns; pure, azure, green, like nothing else in the world, fiery, black; mathematically exact, . . . There are crystal grottos, . . . architecture and engineering art . . . Egypt crystallizes in pyramids and obelisks, Greece in columns; the middle ages in vials; London in grimy cubes . . . To equal nature it is necessary to be mathematically and geometrically exact.

He added a drawing to his words to express his humility in front of these miracles of nature:

Here are some beautiful crystal shapes from the collection of the University of Budapest. These can be seen by the naked eye.

Amethyst, Baiut, Romania

OVERLEAF: *Open-air sculpture depicting the internal structure of a crystal, Seoul, Korea*

Magnetite adularia, Binnenthal, Switzerland

Pyromorphite, Dognacea, Romania

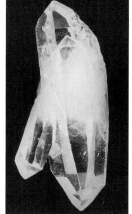

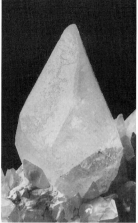

Calcite, Gyöngyösoroszi, Hungary

Amethyst, Telkibánya, Hungary

Calcite, Budapest, Hungary

Other crystals, much smaller, were photographed with an electron microscope. These photos show them at a few hundred to a thousand times magnification.

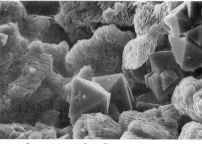

Nesquehonite

Northupite and calcite

The external symmetry of a crystal is due to its internal structure. However, the symmetry of the crystal shape and the symmetry of its internal arrangement may be fundamentally different.

Northupite

Wurtzite

Thenardite

Rotation:
When an object is rotated around its axis, it appears in the same position two or more times

Symmetry element (tool)
Axis of rotation

Reflection:
Reflecting one-half of an object reconstructs the image of the whole object

Symmetry element (tool)
Mirror plane
Applying a mirror plane to either of two halves, the whole is recreated

External Symmetry of Crystals and the Magic Number 32

The external symmetry of crystals can be characterized by reflection planes and rotation axes, similar to the symmetry of polyhedra *(see p. 87),* as the crystal shapes are indeed the shapes of polyhedra.

There are 32 possible crystal shapes. They are called the 32 Crystal Groups. They are shown by examples of actual minerals. (For one of the 32, no mineral has yet been found.)

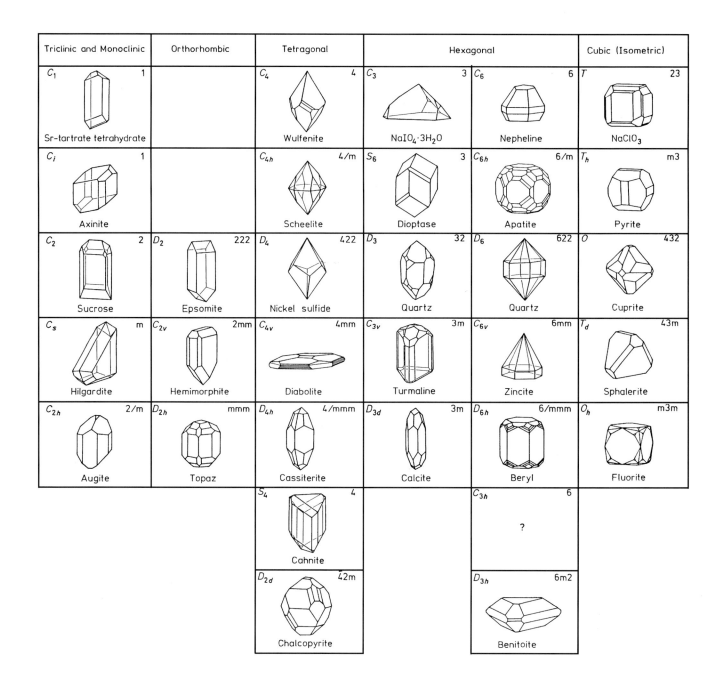

Triclinic and Monoclinic	Orthorhombic	Tetragonal	Hexagonal		Cubic (Isometric)
C_1 1 Sr-tartrate tetrahydrate		C_4 4 Wulfenite	C_3 3 $NaIO_4 \cdot 3H_2O$	C_6 6 Nepheline	T 23 $NaClO_3$
C_i 1̄ Axinite		C_{4h} 4/m Scheelite	S_6 3̄ Dioptase	C_{6h} 6/m Apatite	T_h m3 Pyrite
C_2 2 Sucrose	D_2 222 Epsomite	D_4 422 Nickel sulfide	D_3 32 Quartz	D_6 622 Quartz	O 432 Cuprite
C_s m Hilgardite	C_{2v} 2mm Hemimorphite	C_{4v} 4mm Diabolite	C_{3v} 3m Turmaline	C_{6v} 6mm Zincite	T_d 4̄3m Sphalerite
C_{2h} 2/m Augite	D_{2h} mmm Topaz	D_{4h} 4/mmm Cassiterite	D_{3d} 3̄m Calcite	D_{6h} 6/mmm Beryl	O_h m3m Fluorite
		S_4 4̄ Cahnite		C_{3h} 6̄ ?	
		D_{2d} 4̄2m Chalcopyrite		D_{3h} 6̄m2 Benitoite	

Stereographic Projections

The 32 Crystal Groups can also be represented by so-called stereographic projections. For how these projections are made, see the next page.

Triclinic and Monoclinic	Orthorhombic	Tetragonal	Hexagonal		Cubic (Isometric)
C_1 1		C_4 4	C_3 3	C_6 6	T 23
C_i $\bar{1}$		C_{4h} 4/m	S_6 $\bar{3}$	C_{6h} 6/m	T_h m3
C_2 2	D_2 222	D_4 422	D_3 32	D_6 622	O 432
C_s m	C_{2v} 2mm	C_{4v} 4mm	C_{3v} 3m	C_{6v} 6mm	T_d $\bar{4}$3m
C_{2h} 2/m	D_{2h} mmm	D_{4h} 4/mmm	D_{3d} $\bar{3}$m	D_{6h} 6/mmm	O_h m$\bar{3}$m
		S_4 $\bar{4}$		C_{3h} $\bar{6}$	
		D_{2d} $\bar{4}$2m		D_{3h} $\bar{6}$m2	

201

Preparation of Stereographic Projections

Stereographic projections represent the three-dimensional crystal shapes in two dimensions.

First, draw the polyhedron (corresponding to the crystal shape) and draw a circle around it. This corresponds to a sphere around the crystal. Then extend the face normals (the lines perpendicular to the faces of the polyhedron) to reach the surface of the sphere as seen below. Thus, a set of points representing the faces of the crystal will occur on the surface of the sphere:

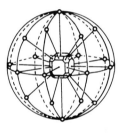

Then draw a line from all the points in the Northern Hemisphere to the South Pole and mark the points on the equatorial plane with filled circles where these connecting lines intersect this plane. This will create a representation of the faces of the upper half of the crystal within a single circle, as seen on the figure (only the front lines and points are indicated on these drawings):

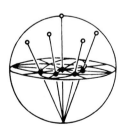

The points that were on the equator originally will remain there:

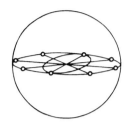

Next, connect the points in the Southern Hemisphere to the North Pole and mark the points on the equatorial plane with open circles:

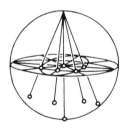

Finally, project the equatorial circle on the plane and indicate all the points that were drawn into the previous figures. Thus, we arrive at a representation of the whole crystal within a circle:

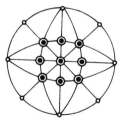

(with all the points indicated here).

The initial and final steps of preparing stereographic projections are given below for the cube and for the octahedron:

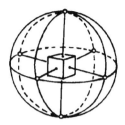

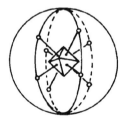

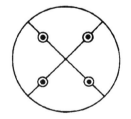

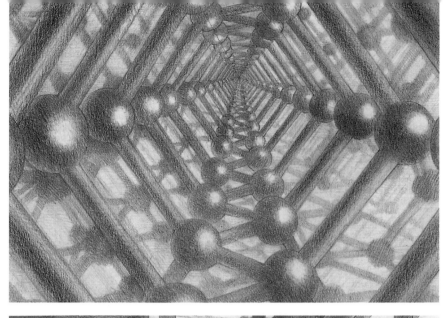

Internal Structure of Crystals

The symmetry of the internal structure of the crystal is characterized by periodicity in three directions. Thus, what we have seen in border decorations in one direction *(see p. 124, for example)* and for planar patterns in two directions *(see p. 184, for example),* we will now see for crystals in three directions.

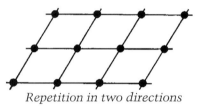

Repetition in one direction

Repetition in two directions

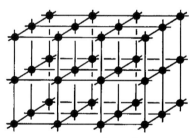

Repetition in three directions (three dimensions)

Two models of three-dimensional crystal lattices, the lower after an M. C. Escher drawing

Periodicity:
The repetitive occurrence of exactly the same motif

Repetition:
Repeated application of the same symmetry operation—simple translation, reflection, or rotation

Symmetry element (tool)
The tool corresponding to whatever operation is repeated (**mirror plane** for reflection, **axis of rotation** for rotation, **constant shift** for translation)

Kwan-Mo Chung, Cosmonergy. *Artist's rendition of internal structure: part of an open-air sculpture at a busy intersection in Seoul, Korea*

XV. DIAMONDS & GLASS

The Magic Number 230

Simple symmetry operations and their combinations provided 7 possibilities for creating border decorations and 17 possibilities for creating planar networks. For three-dimensional periodicity, there are altogether 230 possibilities. The role of the basic motif is played by a small part of the crystal structure, whose "infinite repetition" by the symmetry operations of any of the 230 possibilities produces the entire crystal structure. We can identify a very small part of the crystal as its basic building block. The infinite repetition of this part by symmetry operations builds the whole crystal. This basic building block of the crystal is called the *unit cell*.

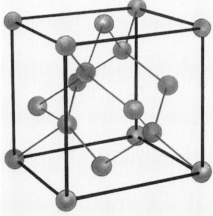

Unit cell of diamond crystal; the spheres represent the carbon atoms

Diamond and Graphite

Diamonds consist solely of carbon. Each carbon atom is surrounded by four other carbon atoms in a tetrahedral configuration. The carbon atoms are linked by strong bonds in all four directions. It is an extremely simple and stable structure, which accounts for the remarkable strength and hardness of diamonds.

Carbon has another crystalline form, graphite.* As we have seen in Chapter XIII *(p. 169)*, graphite has a layer structure and is not nearly as hard as diamond. Curiously, graphite is more stable than diamond under ordinary conditions. If left alone, carbon will form graphite, rather than diamond. To form diamonds artificially, crystallization must be done at very high pressure, tens of thousands of atmospheres, because diamond is much denser than graphite. Under ordinary conditions (that is, room temperature and atmospheric pressure), diamond will eventually turn into graphite. "Eventually" in this case means a very long time indeed. In fact, under ordinary conditions, diamond will last longer than the present age of our universe.

*Until recently, diamond and graphite were thought to be the only modifications of carbon. Today we know of a third one as well, buckminsterfullerene *(see p. 100)*.

Diamond ring

Diamond and Glass

As we have seen, there is a rich variety of symmetry in the crystal forms of minerals. A piece of glass, however, may also be made into a highly symmetrical shape. Even though the piece of glass may acquire the same outer form as a piece of diamond, it will not acquire all the other properties of the diamond. This fact was recognized long ago. In 6th-century India, as portrayed in the *Kama Sutra* by Vatsyayana, one of the arts which a courtesan had to learn was mineralogy. If she were paid in precious stones, she had to be able to distinguish real crystals from paste. With the discovery of X-ray diffraction in 1912 it became possible to determine clearly whether a small, hard crystal was indeed a diamond.

Glass breaks easily. In fact, a diamond knife is often used to cut glass. The difference in strength comes from the difference in internal structure. Glass consists of silicon and oxygen atoms bound together in alternating but random fashion.

Silicon and oxygen atoms can build a crystal, too—such as quartz. As quartz is crystalline, it is characterized by three-dimensional periodicity, whereas glass is amorphous, with no such order in its structure. The difference can be seen by comparing the two structures in a two-dimensional representation, as at right.

To summarize, the fundamental differences in the properties of diamond and glass originate from the difference of symmetries in their internal structures. Diamonds are crystals and there is a highly symmetrical arrangement of carbon atoms in three dimensions. Glass is not a crystal and its internal structure lacks symmetry; there is no regularity and there is no periodicity.

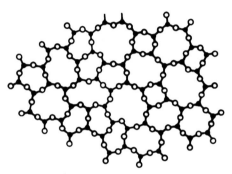
Glass has an amorphous structure (two-dimensional representation)

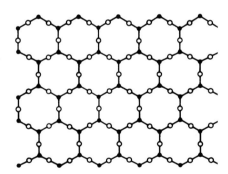
Structure of quartz crystal (two-dimensional representation)

Packing

The building particles of a crystal may be atoms, molecules, or ions. Diamond and graphite consist of carbon atoms, ice consists of water molecules, and common salt (sodium chloride) consists of sodium ions (Na^+) and chloride ions (Cl^-). The internal structure of crystals ensures the densest packing of the building particles. The stick and ball model, often used to depict the internal structure of crystals, is convenient to show the arrangement of the building particles in the crystal but does not convey a realistic impression of densest packing. *(See also p. 113 for the representations of molecules.)*

When Johannes Kepler examined snow crystals, he arrived at the idea of densest packing intuitively.

He did not know about molecules and atoms, but imagined the internal structure of the snowflake as a heap of densely packed balls. He described this in 1611, and his drawing is very much like a heap of cannonballs.

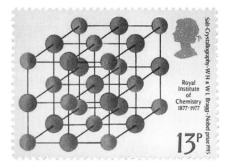

Crystal structure on British stamp

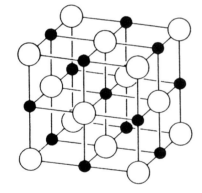

Arrangement of sodium and chloride ions in the common salt crystal

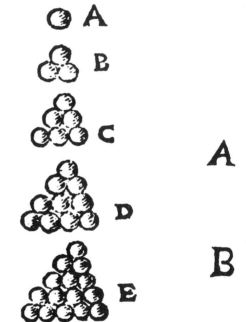

A better model for this purpose is one in which the building particles touch each other:

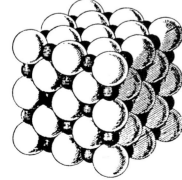

The packing of sodium and chloride ions in the common salt crystal

ABOVE: *Kepler's drawings of the internal structure of snowflakes*
BELOW: *Cannonballs, Laconia, New Hampshire*

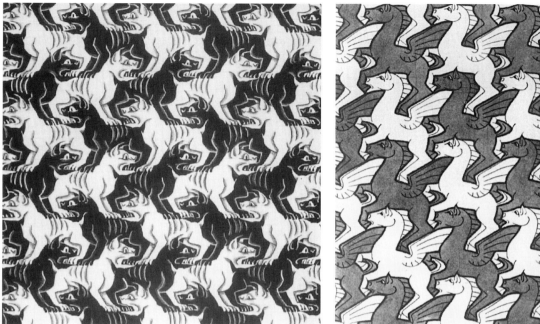

M. C. Escher's Symmetry Drawing E 97—1955   *M. C. Escher's* Symmetry Drawing E 105—1959

If the packing particles are simply spheres, such as carbon atoms or sodium and chloride ions, not much variation is possible. However, when molecules (which are seldom spherical) are building up the crystal, minimizing the empty space between molecules to achieve densest packing is no trivial matter. Because empty space has to be minimized, symmetry planes are rarely present in the internal structure of crystals. Rather, the molecules are arranged in such a way that the concave part of one molecule accommodates the convex part of the other molecule. This is called *dove-tail packing*.

M. C. Escher's periodic drawings of dogs and of winged horses are excellent illustrations of this type of densest packing. Of course, these illustrations are only in two dimensions, and crystals extend in three dimensions, so you have to use your imagination.

For an existing crystal, dove-tail packing is illustrated by an organic molecule with a complicated name (1,3,5-triphenylbenzene).

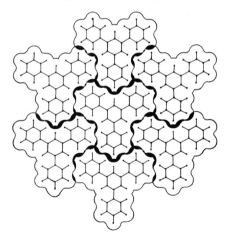

1, 3, 5-triphenylbenzene

Quasicrystals

Quasicrystals are somewhere between amorphous bodies (like glass) and perfect crystals (like diamonds).

Crystals are *regular* and *periodic*. These stipulations set severe limitations on the internal arrangements that can make a crystal. To understand this, we will look again at two-dimensional models. We have seen that it is impossible to cover a surface with equal-size regular pentagons *(see p. 169)*. This can serve to illustrate (in our imagination) that certain symmetries—most notably 5-fold symmetry—are impossible in three-dimensional networks as well. We have also seen—in the example of the regular pentagons—that simple rules may be set up that allow regular pentagons of gradually changing size to cover the available surface. Five-fold symmetry is sometimes called noncrystallographic symmetry, as it is forbidden in the world of crystals. There have been many attempts, at least since Kepler's time, to construct patterns that—at least in a nonperiodic way—display 5-fold symmetry.

Creating a planar pattern with gradually changing-size regular pentagons *(see p. 170)* was originally considered more of a mathematical recreation than a truly scientific endeavor. The most famous planar network with long-range pentagonal regularity was created in tiles by the Oxford mathematician Roger Penrose in the early 1970s.

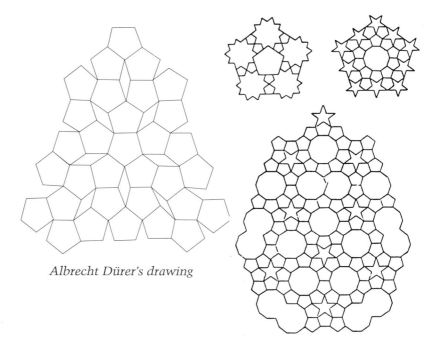

Albrecht Dürer's drawing

Johannes Kepler's drawings

Penrose tiling

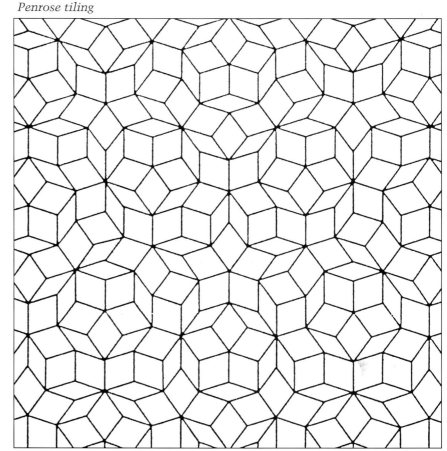

Serendipity

In 1982, Israeli scientist Dan Shechtman unexpectedly discovered the existence of three-dimensional solids with regular and nonperiodic internal structures. Such structures are called *quasicrystals*.

There is no order in glass. There is no way to tell from the structure of one part what another part's structure will be. With diamond, on the other hand, knowing the structure of the unit cell tells us the structure of the entire crystal, and it will be the same throughout.

The symmetry of a quasicrystal is categorized as being somewhere in between the lack of symmetry of amorphous glass and the perfect symmetry of diamond. Although the structure is not the same throughout, there are simple rules that enable us to describe it.

Many scientists were greatly surprised at first by the discovery of quasicrystals. It took Professor Shechtman two years to get the report of his discovery published. The appearance of his paper in 1984 caused a minirevolution in solid state physics and crystallography. Hundreds and hundreds of papers have followed. It is interesting to note that what started as an exercise in symmetry considerations evolved into a new branch of science.

ABOVE: *Dan Shechtman (1991)*

LEFT: *Quasicrystal Al-Li-Cu*

BELOW: *Sculpture resembling a quasicrystal by Swiss sculptor Peter Hächler, Lenzburg, Switzerland*

BELOW: *Scanning electron micrograph of quasicrystalline Al-Cu-Ru*

EPILOGUE

One can only marvel at the richness and diversity in the worlds of symmetry. Yet what we have seen here, in our mostly visual journey, is merely the tip of the iceberg. We have just scratched the surface. All of the subjects introduced, all the photos and drawings, all the roads embarked upon here can lead in many directions and may serve to introduce you to further discoveries and newer insights.

A word of caution, however: at times you may get truly saturated with thoughts of symmetry. Focussing too much attention on regulation, repetition, balance or uniformity can be confining and irritating—can even become obsessive. Perfection may not be a suitable characteristic for human habitat. For example, in the 14th century Japanese *Essays in Idleness*, it is said that: "In everything . . . uniformity is undesirable. Leaving something incomplete makes it interesting, and gives one the feeling that there is room for growth . . . Even when building the imperial palace, they always leave one place unfinished."

Or in *The Magic Mountain*, Thomas Mann writes that snowflakes are " . . . too regular, as substance adapted to life never was to this degree . . ." and how " . . . builders of antiquity purposely and secretly introduced minute variation from absolute symmetry in their columnar structures."

Similarly, some Muslim rug weavers believe that to make a pattern perfect will trap their soul. Therefore they introduce a very small change of color in an otherwise perfect pattern.

Indeed, lack of symmetry may be as appropriate in some cases as its presence may be in others. Nevertheless, we find symmetries everywhere; they are part of our environment and part of our existence.

Let us never forget then, that although symmetry is a fascinating concept for exploration and for unifying diverse fields of human endeavor, it is not symmetry alone that so enriches our lives. It is just an ingredient in nature and human creation, merely a tool. Symmetry only helps us understand and appreciate the beauty and wonder of our universe.

Remember the person mentioned in the introduction who got irritated by the ubiquity of symmetries? We hope that something similar (minus the irritation!) will happen to you as a result of this book, that you will see new patterns and make new connections in the world around you. We hope that the concepts illustrated here will lead you to further exploration and study. And lastly, we hope that this heightened awareness will become a positive and enjoyable factor in your life.

NOTE: If you wish to contribute to a future book on symmetry, write:

István & Magdolna Hargittai
Eötvös University
H-1431 Budapest, Pf. 117 Hungary

FURTHER READING

Bunck, Bryan. *Reality's Mirror: Exploring the Mathematics of Symmetry.* John Wiley & Sons, Inc., New York, 1989

Cook, Theodore Andrea. *The Curves of Life: An Account of Spiral Formations and Their Application to Growth in Nature, to Science and to Art.* Dover Publications, Inc., New York, 1979

Coxeter, H. S. M. *Regular Polytopes.* Third Edition. Dover Publications, Inc., New York, 1979

Feynman, R. *The Character of the Physical Law.* MIT Press, Cambridge, Massachusetts, 1962

Gardner, Martin. *The New Ambidextrous Universe: Symmetry and Asymmetry from Mirror Reflections to Superstrings.* Third Revised Edition. W. H. Freeman and Co., New York, 1990

Grünbaum, Branko, and Shephard, G. C. *Tilings and Patterns.* W. H. Freeman and Co., New York, 1987

Huntley, H. E. *The Divine Proportion: A Study in Mathematical Beauty.* Dover Publications, Inc., New York, 1970

Kapraff, Jay. *Connections: The Geometric Bridge between Art and Science.* McGraw-Hill, Inc., New York, 1991

Loeb, Arthur. *Color and Symmetry.* Wiley-Interscience, New York, 1971

MacGillavry, C. H. *Symmetry Aspects of M. C. Escher's Periodic Drawings.* Bohn, Scheltema and Holkema, Utrecht, Netherlands, 1976

Pearce, Peter, and Pearce, Susan. *Polyhedra Primer.* Dale Seymour Publications, Palo Alto, California, 1978

Rosen, Joe. *Symmetry Discovered: Concepts and Applications in Nature and Science.* Cambridge University Press, Cambridge, U.K., 1975

Schattschneider, Doris. *Visions of Symmetry: Notebooks, Periodic Drawings, and Related Work of M. C. Escher.* W. H. Freeman and Co., New York, 1990

Senechal, Marjorie, and Fleck, George (Editors). *Patterns of Symmetry.* University of Massachusetts Press, Amherst, 1977

Shubnikov, A. V., and Koptsik, V. A. *Symmetry in Science and Art.* Plenum Press, New York, 1972

Stevens, Peter S. *Handbook of Regular Patterns. An Introduction to Symmetry in Two Dimensions.* MIT Press, Cambridge, Massachusetts, 1981

Stewart, Ian, and Golubitsky, Martin. *Fearful Symmetry: Is God a Geometer?* Blackwell Publishers, Oxford, U.K., 1992

Thompson, D'Arcy W. *On Growth and Form.* Cambridge University Press, Cambridge, U.K., 1971

Washburn, Dorothy K., and Crowe, Donald W. *Symmetry and Culture: Theory and Practice of Plane Pattern Analysis.* University of Washington Press, Seattle, 1988

Weyl, Hermann. *Symmetry.* Princeton University Press, Princeton, New Jersey, 1952

Wigner, Eugene P. *Symmetries and Reflections.* Indiana University Press, Bloomington, 1967

Zee, Anthony. *Fearful Symmetry; The Search for Beauty in Modern Physics.* Macmillan Publishing Co., New York, 1986

Other Symmetry Books by the Authors

Hargittai, I., and Hargittai, M. *Symmetry through the Eyes of a Chemist.* Second, revised edition, Plenum Press, New York, 1995

Hargittai, I. (Editor). *Symmetry: Unifying Human Understanding.* Pergamon Press, New York, 1986

———(Editor). *Symmetry 2: Unifying Human Understanding.* Pergamon Press, Oxford, U.K., 1989

———(Editor). *Quasicrystals, Networks, and Molecules of Fivefold Symmetry.* VCH Publishers, Inc., New York, 1990

———(Editor). *Fivefold Symmetry.* World Scientific Publishing Co. Inc., Singapore, 1992

Hargittai, I., and Pickover, C. A. (Editors). *Spiral Symmetry.* World Scientific Publishing Co. Inc., Singapore, 1992

Hargittai, M., and Hargittai, I. *Let's Discover Symmetry* (in Hungarian, for children). Tankönyvkiadó, Budapest, 1989. (An English version is planned.)

ABOUT THE AUTHORS

István and Magdolna Hargittai on a lava slope of a volcano on the Big Island of Hawaii in May, 1993

István Hargittai is Professor of Chemistry and Head of Institute at Budapest Technical University. He is also a member of the Hungarian and Norwegian Academies of Sciences. He attended Eötvös University in Budapest, then got his master's degree in 1965 at Moscow University. He has since worked for the Hungarian Academy of Sciences and later for the Technical University. He created a lab of structural chemistry in Budapest. He is a Ph.D. from Eötvös University, a D.Sc. from the Hungarian Academy of Sciences and a Dr. h.c. from Moscow University. He did postdoctoral work at the University of Oslo and at the University of Texas at Austin.

Since the early '80s he has been a visiting professor at various American universities (Texas, Connecticut, Hawaii) for a total of five years. He continues to work in close cooperation with international scientists and has lectured in 23 countries. He has written over 200 research papers, written and edited 18 books, and served on the editorial board of 12 international journals and book series. He received an Award of Excellence from the Association of American Publishers for one of his symmetry volumes. István developed an interest in photography as a teenager and bought his first camera at age 13, with earnings from tutoring.

Magdolna (Magdi) Hargittai is a science advisor/research professor in the Structural Chemistry Research Group of the Hungarian Academy of Sciences. She got her master's degree in 1969 from Eötvös University and her Ph.D. and D.Sc. degrees from Eötvös University and the Hungarian Academy of Sciences. She has published widely and has been invited to speak in several countries. She has published numerous research papers and has also been involved in writing and editing scientific monographs. She has been the book review editor of *Structural Chemistry*, an international journal. She took up photography as a hobby recently, adding it to her other nonprofessional interests. One is cooking; she published a small Hungarian cookbook in the U.S. in 1986. Her latest project has taken her back to nature: exploration of Hawaiian flora and related legends.

István and Magdi are married and currently live in Budapest, Hungary. They have two children, Balázs, born in 1970, and Eszter, born in 1973. Both Balázs and Eszter spent several years in the U.S., since the family travelled together when István was a visiting professor. Now they are back, but this time on their own. Balázs is a doctoral student in chemistry at Northwestern University in Evanston, Illinois. Although he is in the same field as his parents, his special interest, synthetic organic chemistry, is quite different from his parents' work. Eszter is a sociology major at Smith College in Northampton, Massachusetts. She is also studying media and communications. She worked at Shelter Publications for a week recently and it coincided with the preparation of this book. Thus we had the input of a third Hargittai.

Eszter and Balázs

István and Magdi specialize in molecular structure research. They are interested especially in the simplest, most fundamental molecules. They use an experimental technique called electron diffraction. In an electron diffraction apparatus, very fast electrons are deflected by leading them into the field of molecules. The deflected electrons strengthen and weaken each other (called interference) and the resulting diffraction pattern is photographed.

A typical pattern is shown at right. It consists of a system of concentric rings. They resemble the pattern you see when you throw a stone into still water.

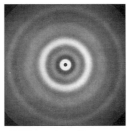

Diffraction diagram

From the diffraction pattern, it is possible to determine the distances between the atoms in the molecule. Then, from the interatomic distances, the shape and symmetry and the entire geometry of the molecule can be reconstructed. The Hargittais, along with colleagues, have built a unique electron diffraction apparatus over the years in Budapest. It is based on a rudimentary apparatus they purchased 25 years ago from a factory in the Ukraine. Their special equipment has attracted visiting scientists to the lab from all over the world.

The symmetry of molecules, alas invisible to the naked eye, was the original incentive that directed the Hargittais' attention to the symmetries of the visible world. They have been involved in the creation of several books on symmetry-related topics, but this is their first attempt at reaching the general public with their experiences and ideas on the subject of symmetry.

INDEX

ACKNOWLEDGEMENTS AND PERMISSIONS

All photos and drawings are by the authors unless noted otherwise

CHAPTER I:

PAGE 1. **Tiger**—*Fedezzük föl a szimmetriát!*, M. Hargittai, I. Hargittai. Tankönyvkiadó, Budapest, 1989

PAGE 2. **Butterflies**—*The Observer's Book of Butterflies*, W. J. Stokoe. © Frederick Warne & Co., London, 1937, 1979

PAGE 3. **Tiger**—*Fedezzük föl a szimmetriát!*, M. Hargittai, I. Hargittai. Tankönyvkiadó, Budapest, 1989

PAGE 4. Background **Redwood leaf**—*Art Forms from Plant Life*, William M. Harlow. Dover Publications, Inc., New York, 1976

PAGE 5. **Buddha**—Ferenc Hopp Museum of Eastern Asiatic Arts, Budapest, photo by Agnes Kolozs; **King Mykerinos**—Lehnert & Landrock, Cairo; **Nude**—Charles Wilp, Düsseldorf; *The Magic Mountain*—This passage is in French in both the original German and English translation. English translation provided by Dr. Jack M. Davis, Professor of English, University of Connecticut, Storrs, Connecticut

PAGE 6. **Gymnast**—Tamás Szigeti, Magyar Hírlap, Budapest; **Swimmer**—MTI-Fotó Archive, Budapest; **Pushups, parallel bars, diver**—Budd Symes Photography, Los Angeles

PAGE 7. **Aztec**—*American Indian Design & Decoration*, Le Roy H. Appleton. Dover Publications, Inc., New York, 1971; **Hungarian king**—*Fedezzük föl a szimmetriát!*, M. Hargittai, I. Hargittai. Tankönyvkiadó, Budapest, 1989

PAGE 8. **Masks**—*American Indian Design & Decoration*, Le Roy H. Appleton. Dover Publications, Inc., New York, 1971; and *Authentic Indian Designs*, Edited by Maria Naylor. Dover Publications, Inc., New York, 1975

PAGE 9. **Picasso**—Museum of Fine Arts, Budapest; **Edgar Allan Poe**—American Antiquarian Society, Worcester, Massachusetts

PAGE 10. **Leuven**—Anne Watteyne, Leuven, Belgium

PAGE 12. **Heros' Square**—János Váraljai, MTI-Fotó, Budapest

PAGE 14. **William Blake**—Fitzwilliam Museum, Cambridge, U.K.; **Children's drawings**—Komló Music School, Komló, Hungary, courtesy of Mária Apagyi, Pécs, Hungary

CHAPTER II:

PAGE 15. **Parachute**—*Fedezzük föl a szimmetriát!*, M. Hargittai, I. Hargittai. Tankönyvkiadó, Budapest, 1989 *(digitally manipulated)*

PAGE 18. **Bugs** (stamps excluded)—*Animals*, selected by Jim Harter. Dover Publications, Inc., New York, 1979

PAGE 20. **Work truck, boats**—Dr. István Gera, Budapest; **Trolley bus, tram**—*Fedezzük föl a szimmetriát!*, M. Hargittai, I. Hargittai. Tankönyvkiadó, Budapest, 1989

PAGE 21. **BMW**—Dr. István Gera, Budapest; **Airship, motorcycle**—*Fedezzük föl a szimmetriát!*, M. Hargittai, I. Hargittai. Tankönyvkiadó, Budapest, 1989

PAGE 22. *Apollo 11, Apollo 9*—NASA, Washington, D.C.; **Hot-air balloons**—AP/Wide World Photos, New York

PAGE 23. **Pollen**—Dr. R. Klockenkämper, Institute für Spektrochemie, Dortmund, Germany

PAGE 24. **Tree near Aveley**—C.T. Ballard; **Tree rings**—*Art Forms from Plant Life*, William M. Harlow. Dover Publications, Inc., New York, 1976

PAGE 25. **Mt. Fuji**—AP/Wide World Photos, New York; **Mushroom, mushroom cloud**—*Fedezzük föl a szimmetriát!*, M. Hargittai, I. Hargittai. Tankönyvkiadó, Budapest, 1989

PAGE 26. **Salt columns**—Palphot Ltd., Herzlia, Israel; **Copper**—Dr. Maria Kazinets, University of Be'er Sheva', Israel;

Iron—Dr. J. Morral, University of Connecticut, Storrs, Connecticut

CHAPTER III:

PAGE 28. **Hands**—Lloyd Kahn, Bolinas, California

PAGE 29. *I Would Like to Be Loved*—Vera Székely, Mulleron-Janury, France

PAGE 30. **Covers of** *Der Spiegel*— June 15, 1992, "Farewell to God" and May 18, 1992, "Who Owns the Earth," Spiegel Publications, Hamburg

PAGE 31. **Soccer player**—Budd Symes Photography, Los Angeles

PAGE 32. **Amino acid**—*The Galactic Club: Intelligent Life in Outer Space*, © 1975 by Ronald N. Bracewell. Reprinted by permission of W. H. Freeman and Co., New York

PAGE 33. **Spiral staircase**—*The Origins of Life: Molecules and Natural Selection*, L. E. Orgel. John Wiley & Sons, Inc., New York, 1973

PAGE 34. **Computer illustrations**—Suzanne Parks, Bolinas, California, based on drawings by Ferenc Lantos, Pécs, Hungary; **Apples**—Lloyd Kahn, Bolinas, California

CHAPTER IV:

PAGE 38. **Computer illustrations**—Suzanne Parks, Bolinas, California, after drawings by Ferenc Lantos, Pécs, Hungary

PAGE 39. **Making pinwheel**—Ferenc Lantos, Pécs, Hungary

PAGE 40. **Water windmill**—*A Field Guide to American Windmills*, by T. Lindsay Baker. © 1985 by the University of Oklahoma Press, Norman, Oklahoma

PAGE 41. **Windfarm**—George Steinmetz. Kenetech/U.S Windpower, Inc., Oakland, California; **Pelton wheel**—Lloyd Kahn, Bolinas, California

PAGE 45. **Blut und Eisen**—John Heartfield, Akademie der Künste zu Berlin, represented by VG Bild-Kunst, Bonn, Germany, 1993, photo: B. Kuhnert; **Jellyfish**—*Art Forms in Nature*, Ernst Häckel. Dover Publications, Inc., New York, 1974

PAGE 49. **Folk art**—*American Indian Design & Decoration*, Le Roy H. Appleton. Dover Publications, Inc., New York, 1971; and *Authentic Indian Designs*, Edited by Maria Naylor. Dover Publications, Inc., New York, 1975

PAGE 50. **Spirograph drawings**—Balázs Hargittai, Evanston, Illinois

CHAPTER V:

PAGE 53. **Computer illustrations**—Suzanne Parks, Bolinas, California, after drawings by Ferenc Lantos, Pécs, Hungary

PAGE 57. **Oriental poppies**—Lloyd Kahn, Bolinas, California

PAGES 58–59. **Jellyfish, starfish, star corals and sea urchins**—*Art Forms in Nature*, Ernst Häckel. Dover Publications, Inc., New York, 1974

PAGE 59. **Starfish with 11 legs**—*Fedezzük föl a szimmetriát!*, M. Hargittai, I. Hargittai. Tankönyvkiadó, Budapest, 1989

PAGE 61. **Drawings**—Ferenc Lantos, Pécs, Hungary

PAGE 66. **Qutb Minar**—*Geometric Concepts in Islamic Art*, Issam El-Said & Ayse Parman. World of Islam Festival Trust Publishing Co., London, 1976

PAGE 68. **Six papercuts**—Eszter Hargittai, Northampton, Massachusetts; **Two Chinese papercuts**—*The Art of Chinese Papercuts*, Zhang Daoyi. Foreign Languages Press, Beijing, 1989

Chapter VI:

Pages 69–78. **Snowflakes**—*Snow Crystals*, W. A. Bentley and W. J. Humphreys. Dover Publications, Inc., New York, 1962

Page 71. **Ice crystal**—Reprinted from *The Nature of the Chemical Bond*, 3rd Edition, Linus Pauling © 1960 by Cornell University. Used by permission of the publisher, Cornell University Press, Ithaca, New York

Page 73. **Yin/Yang poem**—*Weather*, Vol. 16, p. 319, J. Needham & Lu Gwei-Djen, 1961

Page 74. **Descartes illustration**—*Snow*, U. Nakaya. Iwanami-Shoten Publishers Co., Tokyo, 1938

Pages 74–75. **Scoresby sketches**—*Arctic Scientist*, William Scoresby. Caedmon of Whitby Publishers, U.K., 1976

Page 75. **Sekka Zusetsu sketches**—*Snow*, U. Nakaya. Iwanami-Shoten Publishers Co., Tokyo, 1938

Page 78. **Photomicrograph and sketches**—*Snow*, U. Nakaya. Iwamani-Shoten Publishers Co., Tokyo, 1938

Chapter VII:

Page 80. **Buildings from above**—*Fedezzük föl a szimmetriát!*, M. Hargittai, I. Hargittai. Tankönyvkiadó, Budapest, 1989

Page 82. **Washington Monument**—AP/Wide World Photos, New York

Page 83. **Castillo de San Marcos**—National Park Service, Castillo de San Marcos, St. Augustine, Florida; **Goryokaku Castle**—Hakodale City office, Hokkaido, Japan; **Pentagon**—AP/Wide World Photos, New York

Chapter VIII:

Page 94. **Radiolarians**—*Kunstformen der Natur*, Vols. 1–10, Ernst Häckel. Verlag des Bibliographischen Instituts, Leipzig, Germany, 1899–1904; **Polyoma virus**—Drawn after K. W. Adolph, D. L. D. Caspar, C. J. Hollingshed, E. E. Lattman, W. C. Phillips, and W. T. Murakami, *Science*, Vol. 203, p. 1117, 1979; **Packing of spheres**—Drawn after A. L. Mackay, *Acta Crystallographica*, Vol. 15, p. 916, 1962

Page 97. **S. Dali**—Drawing by Ferenc Lantos, Pécs, Hungary, after a photograph; **Crystal Slave**—Horst Janssen, Christians Verlag, Hamburg; **Stained glass models**—Dr. Herbert Hauptman, Buffalo, New York

Page 99. **Archimedean polyhedra**—*Polyhedra Primer*, Peter Pearce and Susan Pearce. Dale Seymour Publications, Palo Alto, California, 1978

Page 100. **R. Buckminster Fuller, Expo dome**—Lloyd Kahn, Bolinas, California

Page 101. *Nature*, Macmillan Magazines Ltd., Vol. 318, No. 6024, November 14–20, 1985; *Science*, cover from J. Bernholc. Molecule of the Year cover, Vol. 254, December 20, 1991. Copyright 1991 by AAAS; *New Scientist*, Osnat Lippa, freelance illustrator, July 6, 1991, No. 1776; *Angewandte Chemie*, International Edition in English, Vol. 31, No. 2, 1992, VCH Verlagsgesellschaft, Weinheim, Germany

Page 103. **Earth sphere**—NASA, Washington, D.C.

Chapter IX:

Pages 108–110. **Balloons**—Lloyd Kahn, Bolinas, California

Page 112. **Six walnuts**—Dr. András Határ, Budapest; **Chestnuts**—Dr. Anna Rita Campanelli, Rome

Chapter X:

Page 117. **Eastern Orthodox church**—Dr. Arkadii A. Ivanov, Moscow

Page 118. **Drawings with antirotational symmetry**—*Symmetry and Antisymmetry of Finite Figures* (in Russian), A. V. Shubnikov. Akad. Nauk. S.S.S.R., Moscow 1951; **Yin/Yang**—Suzanne Parks, Bolinas, California

Page 119. **Vasarely print**—Victor Vasarely, Annet-sur-Marne, France

Chapter XI:

Page 124. **Border decorations**—*Japanese Border Designs*, Selected and Edited by Theodore Menten. Dover Publications, Inc., New York, 1975; *Persian Designs and Motifs for Artists and Craftsmen*, Ali Dowlatshahi. Dover Publications, Inc., New York, 1979

Page 125. ***On Montmartre***—Vincent van Gogh. © 1992, The Art Institute of Chicago. All rights reserved

Page 133. **Footsteps**—Drawing by Ferenc Lantos, Pécs, Hungary

Page 134. **Rowers**—Budd Symes Photography, Los Angeles

Page 135. **Hungarian needlework**—Györgyi Lengyel, Budapest

Pages 136–137. **Greek, Roman, Egyptian designs**—*Designs of the Ancient World*, Hart Picture Archives, compiled by R. Sietsema. Hart Publications, New York, 1978

Page 138. **Mexican patterns**—*Design Motifs of Ancient Mexico*, Jorge Enciso. Dover Publications, Inc., New York, 1953; **Onandaga belt**—*Authentic Indian Designs*, Edited by Maria Naylor. Dover Publications, Inc., New York, 1975; **Pottery**—*American Indian Design & Decoration*, Le Roy H. Appleton. Dover Publications, Inc., New York, 1971

Page 139. **Arabic patterns**—*Arabic Art in Color*, Edited by Prisse D'Avennes. Dover Publications, Inc., New York, 1978

Page 140. **Persian designs**—*Persian Designs and Motifs for Artists and Craftsmen*, Ali Dowlatshahi. Dover Publications, Inc., New York, 1979

Page 141. **Japanese designs**—*Japanese Border Designs*, Selected and Edited by Theodore Menten. Dover Publications, Inc., New York, 1975

Page 142. **Chinese designs**—*Chinese Lattice Designs*, Daniel Sheets Dye. Dover Publications, Inc., New York, 1974

Page 144. **Drawing for papercutting**—Ferenc Lantos, Pécs, Hungary; **Border decorations**—*Symmetry and Antisymmetry of Finite Figures* (in Russian), A. V. Shubnikov. Akad. Nauk. S.S.S.R., Moscow, 1951

Chapter XII:

Page 147. **Guggenheim Museum**—Solomon R. Guggenheim Museum, New York; photo: David Heald © 1992, The Solomon R. Guggenheim Foundation, New York; **Impossible stairway**—idea after a movie poster for *Glück im Hinterhaus*

Page 148. **Helical biological macromolecule**—P. Doty in *The Molecular Basis of Life*, Edited by R. H. Haynes & P. C. Hanewalt. W. H. Freeman & Co., San Francisco and London, 1968; **Left- and** right-handed helices—*Biochemistry: A Problems Approach*, W. B. Wood, J. H. Wilson, R. M. Benbow, L. E. Hood. Benjamin Cummings Publishing Co., Menlo Park, California, 1981

PAGE 149. **Hurricane Pefa & Tropical Storm Sam**—NASA, Washington, D.C.; **Whirlpool Galaxy**—UCO/Lick Observatory, University of California, Santa Cruz, California; **Bathtub vortex**—Merwin Sibulkin. *American Scientist*, 1983; **Clouds**—*Die Spirale*, Museum für Gestaltung, Basel, Switzerland (out of print)

PAGE 150. **William Blake**—*Die Spirale*, Museum für Gestaltung, Basel, Switzerland (out of print); **Friedensreich Hundertwasser**— 552 *The Neighbours II: Spiral Sun and Moon House*, 1963, Joram Harel Management, Vienna; **Computer drawings**—*Computers, Pattern, Chaos, and Beauty*, C. Pickover. © 1990 St. Martin's Press, New York. All rights reserved

PAGE 151. **Solarium**—*Animals*, Selected by Jim Harter. Dover Publications, Inc., New York, 1979; **California snail**—Lloyd Kahn, Bolinas, California

PAGE 152. **Goat**—*The Curves of Life*, Theodore Andrea Cook. Dover Publications, Inc., New York, 1979; **Greater kudu**— *Animals*, Selected by Jim Harter. Dover Publications, Inc., New York, 1979; **Bighorn sheep**—Dr. Fred Lipschultz, Storrs, Connecticut; **Wild cucumber, storksbill**—*Art Forms from Plant Life*, William M. Harlow. Dover Publications, Inc., New York, 1976

PAGE 153. **Malwiya**—*Geometric Concepts in Islamic Art*, Issam El-Said & Ayse Parman. World of Islam Festival Publishing Company, Ltd., London, 1976; **Galilei Tower**—Caspar Schwabe, Zürich

PAGE 154. **Drawings**—Ferenc Lantos, Pécs, Hungary

PAGE 158. **Sunflowers & cauliflower**—Sándor Kabai, Budapest; *New Scientist* **cover**—*Daisy* by Robert Dixon. *New Scientist*, 17 December 1981, Vol. 92, No. 1284; **Cobweb thistle**—Lloyd Kahn, Bolinas, California

PAGE 159. **Michelangelo's design**—George Bain/artist. © 1990/91 Stuart Titles Ltd., Glasgow, Scotland; **White Mountain Apache**—*American Indian Design & Decoration*, Le Roy H. Appleton. Dover Publications, Inc., New York, 1971

PAGE 163. **Nautilus shell**—Lloyd Kahn, Bolinas, California

PAGE 164. **Proportions of *Adam's Creation***—Drawing by György Doczi, Seattle, Washington

PAGE 166. **Korean designs**—*Korean Motifs: 1 Geometric Patterns*, Edited by Sang-Soo Ahn. Ahn Graphics & Book Publishers, Seoul, 1986

CHAPTER XIII:

PAGE 167. **Newly born bees, Worker bees** (lower left)—*Fedezzük föl a szimmetriát!*, M. Hargittai, I. Hargittai. Tankönyvkiadó, Budapest, 1989; **Worker bees** (upper right)—*The Golden Throng* by Edwin Way Teale. Robert Hale Ltd., London, 1942; **Wax comb**—Zoltán Őrösi

PAGE 168. **Basalt joints**—*Principles of Physical Geology*, A. Holmes. The Ronald Press Company, New York, 1965; **Moth's compound eye**—J. Morral, University of Connecticut, Storrs, Connecticut; **Oil platforms**—Report of the Statoil Co., Norway, 1979

PAGE 169. **Needlework**—Györgyi Lengyel, Budapest

PAGE 170. **Tiling with regular pentagons**—after Professor A. L. Mackay, University of London; **Pentagonal snowflake**— Computer drawing by Robert Mackay, London

CHAPTER XIV:

PAGE 171. **Sophia Loren**—Jet-Set, Denis Taranto, Paris

PAGE 173. **Textile designs**—*Architexture*, Wolfgang Hageney. Edition Belvedere Co., Ltd., Rome-Milan, 1981; **Javanese batik designs**—Annagret Haake, University of Frankfurt

PAGE 175. ***Campbell's Soup Cans & Coca Cola Bottles***—© 1993 Andy Warhol Foundation for the Visual Arts/ARS, New York

PAGE 181. **Polar bear & maple leaf patterns**—Professor François Brisse, *Canadian Mineralogist*, Vol. 19, p. 217, 1981

PAGE 182. **Shibām-Kawkaban, Topkapi Palace, Sidi bu Medien**— *Geometric Concepts in Islamic Art*, Issam El-Said & Ayse Parman. World of Islam Festival Publishing Company, Ltd., London, 1976; **Badra decoration**—Professor Khudu Mamedov, Baku, Azerbaijan

PAGE 183. **Arabic wall mosaics**—*Arabic Art in Color*, Edited by Prisse D'Avennes. Dover Publications, Inc., New York, 1978

PAGES 184–187. **Needlework designs**—Györgyi Lengyel, Budapest

PAGE 188. **Indian bags & baskets**—*Authentic Indian Designs*, Edited by Maria Naylor. Dover Publications, Inc., New York, 1975

PAGE 189. **Persian designs**—*Persian Designs and Motifs for Artists and Craftsmen*, Ali Dowlatshahi. Dover Publications, Inc., New York, 1979; **Korean designs**—*Korean Motifs 1: Geometric Patterns*, Edited by Sang-Soo Ahn. Ahn Graphics & Book Publishers, Seoul, 1986

PAGES 190–191. **Girls, Unity & Seagulls**—Professor Khudu Mamedov, Baku, Azerbaijan

PAGE 191. **Fish & boat**—M. C. Escher, *Symmetry Drawing E 113 (1962)*. © 1962 M. C. Escher Foundation ® – Baarn, Holland. All rights reserved

PAGE 192. **Fish & bird**—M. C. Escher, *Symmetry Drawing E 115 (1963)*. © 1963 M. C. Escher Foundation ® – Baarn, Holland. All rights reserved; **Falcon, fly, butterfly & bat**—M. C. Escher, *Symmetry Drawing E 81 (1950)*. © 1950 M. C. Escher Foundation ® – Baarn, Holland. All rights reserved

PAGE 193. **El Greco, *Study head***—Museum of Fine Arts, Budapest, photo by Dénes Józsa; ***Homage to Greco***—Museum of Fine Arts, Budapest, photo by László Haris; **Elephants**—*Going Going Gone*, African Wildlife Foundation, Washington, D.C.

PAGE 194. **Illustrations**—Drawn after Owen Jones; **Confusion illustration**—*Symmetry and Antisymmetry of Finite Figures* (in Russian), A. V. Shubnikov. Akad. Nauk. S.S.S.R., Moscow, 1951

CHAPTER XV:

PAGE 198. **Crystal cartoon and quotation**—*Anglicke Listy*, Karel Čapek, Ceskoslovensky Spisovatel, Praha, 1970

PAGES 198–199. **Crystals**—Mineral collection of the University of Budapest. Photos by István Gatter and Attila Gömbös, Budapest

PAGE 199. **Electron microscope photographs**—István Dódony, Budapest

PAGES 200–201. **Crystal shapes and projections**—Drawn after *Elementary Crystallography: An Introduction to the Fundamental Geometrical Features of Crystals*, M. J. Buerger. John Wiley & Sons, Inc., New York, 1967; *A Textbook of Mineralogy*, E. S. Dana. John Wiley & Sons, Inc., New York, 1932; *Architecture of Crystals* (in Russian), P. M. Zorkii. Nauka, Moscow, 1968

PAGE 203. **Crystal structures**—Drawings by Ferenc Lantos, Pécs, Hungary

PAGES 204–205. **Photos of diamonds**—Ernst A. Heiniger. *The Great Book of Jewels,* Ernst A. and Jean Heiniger. Edita S. A., Lausanne, Switzerland, 1974

PAGE 206. **Sodium chloride space-filling model**—After W. Barlow in *Zeitschrift für Kristallographie,* Vol. 29, p. 433, 1898

PAGE 207. **Dogs**—M. C. Escher, *Symmetry Drawing E 97 (1955).* © 1955 M. C. Escher Foundation ® – Baarn, Holland. All rights reserved; **Pegasus**—M. C. Escher. *Symmetry Drawing E 105 (1959).* © 1959 M. C. Escher Foundation ® – Baarn, Holland. All rights reserved

PAGE 208. **Kepler's drawings**—L. Danzer, B. Grünbaum and G. C. Shephard, *American Mathematical Monthly,* Vol. 89, pp. 568–585, 1982; **Dürer's drawing**—Donald W. Crowe in *Fivefold Symmetry,* Edited by I. Hargittai. World Scientific, Singapore, 1992

PAGE 209. **Al-Cu-Ru quasicrystal**—Hans-Ude Nissen, Zürich; **Al-Li-Cu quasicrystal**—Françoise Dénoyer, Orsay, France

PAGE 212. **István and Magdolna**—Balázs Hargittai, Evanston, Illinois

SPECIAL THANKS to Dover Publications, Inc., New York, for its remarkable list of reference books.

BOOK CREDITS

EDITOR: Lloyd Kahn

PROJECT MANAGER: Christina Reski

DESIGN, PRODUCTION: Suzanne Parks

TYPESETTING, PRODUCTION: Christina Reski

CONTRIBUTING EDITOR: Stuart Kenter

DESIGN CONSULTANT: David Wills

PROOFREADER, INDEXER: Betty Berenson

QUALITY CHECK: Marianne Rogoff

PHOTO LAB: Chong Lee, San Francisco, California

INTERNEGATIVES: General Graphics Services, San Francisco, California

TECHNICAL SUPPORT: James Mellard, Mellard Digital Arts, Sausalito, California

PRINT PRODUCTION COORDINATOR: Charlie Lane, Quebecor Printing

CUSTOMER SERVICE REPRESENTATIVE: Linda Kyle, Quebecor Printing

IN-HOUSE PRODUCTION HARDWARE: Macintosh FX 20/200; IIcx 8/200; IIci 20/120, 50mhz Daystar acceleration and Supermac 8/24 PDQ video; Fujitsu Optical Drive; Microtek 600zs scanner; Apple Laserwriter Pro 600

SCANNING, LINOTRONIC FILM OUTPUT: Eric Neylon and Victor Milianti, Marinstat Graphic Arts, Inc., Mill Valley, California

SCANNING HARDWARE: Agfa Arcus® 1200 DPI flatbed scanner, Leaf 45 slide scanner

SCANNING SOFTWARE: Adobe Photoshop® 2.5.1

PAGE LAYOUT SOFTWARE: QuarkXpress® 3.2

ILLUSTRATION SOFTWARE: Adobe Illustrator® 5.0

TEXT TYPE: Adobe Frutiger, Adobe Trump Medieval

PAPER, TEXT: 70 lb Warren Dull
PAPER, COVER: 100 lb Lustro Dull

SPECIAL THANKS TO: Steve Baer, for bringing together authors and publisher

AND TO:
Joan Creed
Michele M. Donahue, Allan & Gray Corp.
Dover Publications
Louie Frazier
Eszter Hargittai
Lesley Kahn
Frank. L. Lambert
Ferenc Lantos
George Young

PRESS: Quebecor Printing, Kingsport, Tennessee

PRINTING: 77" Harris Sheetfed Press

Press crew, first printing, symmetrical pose. Clockwise from lower left: James Cassell, Mike Cox, Larry Bradley, Eddie Shepherd